The Transsexual Delusion

G. Eugene Pichler

The Transsexual Delusion

G. Eugene Pichler

THE MEN & WOMEN WHO TRANSGRESS GENDER NORMS

The Transsexual Delusion
The Men & Women Who Transgress Gender Norms
by G. Eugene Pichler

with contributions by Tim Ayerst, M.A. and Bob Heggie

Visit the following URL to forward all inquiries.
http://www.transgression.com

Manuscript Version 1.44

ISBN: 978-1-365-23724-9

Printed in the United States of America.

Dedication

This book is also dedicated to Professor Richard Dawkins of Oxford University, who taught us to question that which we otherwise accept as fact through faith without evidence while abandoning reason and our innate gift of sensibility.

Contents

Preface

In his book, The Transsexual Delusion—the Men and Women, Who Transgress Gender Norms, G. Eugene Pichler looks at transsexualism as a behavioral addiction, much like a sex addiction, Internet addiction or gambling addiction, which is caused by a faulty reward system in the human brain. Pichler further reports that the medications that effectively treat behavioral addictions show good results in treating transsexualism. In fact modern research shows that the drugs that effectively treat epilepsy treat behavioral addictions. Further, a study by Jan Wålinder [Göteborg, Sweden], dated 1967, shows that epilepsy (which otherwise occurs in 0.82% of the population of the G12 nations) is effectively absent in subjects, who exhibit cross-gender proclivities and who are successfully treated for epileptic seizures, according to 43 case studies, dating back to 1954.

Pichler accepts Dr. Ray Blanchard theories into autogynephilia as the explanation in the early stages of the vast majority of male-to-female transsexual narratives (both heterosexual and asexual), but Pichler is not persuaded that Blanchard's theories explain the persistent addictive behavior long after the individual is effectively castrated. (Dr. Ray Blanchard coined the term, autogynephilia, to denote a male's paraphilic tendency to be sexually aroused by the thought or image of himself as a woman.)

In the case of a male-to-female transsexual the more prevalent motivator that remains years after undergoing an initial invasive procedure that protracts the testes is a behavioral addiction (which is firmly established) and not an autogynephilic disorder. The implication for medical practitioners is simple. Break the behavioral addiction, underlying transsexualism, and you break the disorder.

In his book Pichler recommends health jurisdictions divorce themselves from the Harry Benjamin International Gender Dysphoria Association (HBIGDA) and affect sweeping and dramatic changes to the Standards of Care, requiring transsexual candidates, who seek a GRS procedure, to undergo non-invasive therapy, involving medications that directly deal with behavioral addictions for at least 90 days prior to undergoing hormone replacement therapy (HRT).

Given the dramatically high incidence of male-to-female transsexuals, who not only fail, but fail miserably to benefit from invasive transsexual procedures, Pichler recommends that health jurisdictions order rogue practitioners and third-wave health care providers to cease all medication protocols to their transsexual patients, who seek a GRS procedure, and refer these patients to the centralized gender identity clinics. Pichler also recommends that health jurisdiction remove all personnel at the centralized gender identity clinics, who possess a transsexual background, from their positions of authority over transsexual care and replace them with arm's length clinicians. Pichler further recommends that health jurisdictions expand the services of these centralized gender identity clinics to deal with the inevitable influx of new patients.

This book includes a number of case studies of male-to-female transsexuals, who were effectively treated for transsexualism with medications normally prescribed for epilepsy in lieu of invasive procedures, like GRS procedures and orchiectomies.

The book is not written exclusively for a transsexual audience, but rather it is written for anyone with an interest in the transsexual phenomenon.

Special thanks to Maxine Petersen of the Centre for Addiction and Mental Health (CAMH), Rupert Raj of the Sherbourne Health Centre, Dr. Keith Loukes, M.D. of the Sherbourne Health Centre, Dr. Frank Emil Cashman, M.D. of the St. Michael's Hospital, Dr. J. Paul Federoff, M.D. of the Royal Ottawa Hospital, Steven of the University of Toronto, Kevin Hoop, Daphne, Dale O'Leary for articulating The "Transsexual" Delusion in 2011 on his WordPress BLOG, male-to-female transsexual Kassandra and male-to-female transsexual Michaela.

* * * * *

In his book, G. Eugene Pichler reports on the lives of a number of post-operative, male-to-female transsexuals, who failed to benefit from a gender reassignment surgical procedure and watched their respective lives get measurably worse and not better. Pichler also touches on the current psychiatric understanding of transsexual behavior, the underlying causes and the more recent research and theories that bring into question the current understanding. Pichler also discusses transsexualism as an emerging political movement and the LGBT organizations that routinely intimidate and harass medical practitioners, politicians and journalists, when these parties operate outside a prescribed belief system. Pichler also touches on feminism and the understanding of transsexualism through the multi-faceted feminist perspective. Pichler also touches on a number of other sexual phenomena, including cross-dressing, she-males, and the sexual interest towards transvestites, transsexuals and she-males, and how the lines between the various phenomena are blurred. Pichler also reports on effective treatments for transsexualism that do not involve invasive surgery. To base his arguments Pichler compares transsexualism—which is a behavioral addiction kicked off from a sexual response to the consideration of a change in sex—with epilepsy and shows how the drugs and medications that treat epileptic seizures and behavioral additions have good results in treating transsexualism. In fact a study by Jan Wålinder found that the phenomenon is virtually non-existent in persons who are successfully treated for epilepsy. Pichler further shows that if the medical community were to treat transsexualism in the context of an auto-erotic behavioral addiction in lieu of a gender identity disorder, that transsexualism as it is known today would be virtually eradicated within one generation.

* * * * *

As he introduces each subject Pichler brings the reader into a day-in-the-life of the individual—a bad day—but in the vast majority of cases of five-year post-operative transsexuals sadly a typical day. In the case of male-to-female transsexual, Randy Pallister a.k.a., Jennifer Claire Pallister, Pichler brings the reader to Pallister's final art showing at the OHHO Gallery, prior to his death due to substance abuse. In the case of she-male Escort/Actor Rodney Arsenault, a.k.a., Nina Arsenault, a.k.a., Shemale Escort Amber, who reportedly has an IQ of over 145, Pichler brings the reader along to watch Rodney as he is called up to play the humiliating principal role of a transsexual bimbo waitress on the set of CTV Comedy's show, PUNCHED UP. In the case of DRAG politician, Enza Anderson, Pichler brings the reader inside Anderson's janitorial job at Woody's, cleaning up after Kyle Rae's triumphant celebration over the "Supermodel" in municipal elections.

CHAPTER 01

Introduction

It is 5:30 a.m. in the morning in a quiet estate in Montreal, Quebec. There is a glow of light in the sky towards the East. The radio from an alarm clock in room 3 plays. 'Raven'— unfazed by the alarm—is already awake. Raven (not his real name) is a male-to-female transsexual. This is a big day for Raven. Raven is scheduled to undergo what is known as a gender reassignment surgical (GRS) procedure. Raven believes the surgery will correct what he otherwise regards as a birth defect. After undergoing the procedure Raven believes he will effectively take a major step towards womanhood, but not necessarily the final step. [1] After the surgery, Raven plans to undergo speech therapy and further electrolysis.

Raven was born male. Prior to his gender transition, which reportedly began at age 19, Raven reported that he has always felt that he was a woman trapped in a male's body. Raven is one of thousands of men in North America, who are diagnosed as having a gender identity disorder, each year. The American Psychiatric Association regards Gender Identity Disorder (GID) as a major mental illness. In fact people diagnosed with Gender Identity Disorder typically report that they feel discomfort by their assigned birth gender and they desire to be a member of the opposite sex. Raven, himself, reported that at one time he held a razor blade against his wrists with the intent of committing suicide, stemming from a disquiet revolving around gender.

Raven is scheduled to undergo the procedure from plastic surgeon Dr. Pierre Brassard, who operates a private clinic in Montreal, Quebec for people who seek to change their outward gender appearance through surgical means, like Raven. In addition to gender reassignment surgical procedures (or what is medically referred to the combination of a penectomy coupled with a vaginoplasty procedure in the case of male patients) Brassard offers breast augmentation (for males), tracheal shaves (again for males), and phalloplasty, e.g., a skin graft to artificially create a male penis (for women).

The surgical procedure Raven is about to undergo is radical. The procedure involves the amputation of otherwise healthy anatomy and the construction of what is known as a "neo vagina", using the skin of the male penis. The surgery is irreversible. For that reason the psychiatric community only approves the surgery for people like Raven, who have successfully lived as members of the opposite sex in society for at least one year and who reportedly have no ancillary mental disorders. Raven has to have letters from one psychiatrist and one psychologist testifying that he is a good candidate to undergo the procedure. The letter from the psychiatrist is the more important of the two. The psychiatrist must be recognized as a subject matter expert in gender identity disorder.

Kassandra, who underwent the surgical procedure from Dr. Brassard two years before Raven, reported that he received one of his two letters from a psychologist, who wanted to 'break into the field'. Kassandra also reported that a number of the other candidates at the Montreal estate, who underwent the procedure with Dr. Brassard, confided in him that they were disingenuous with the medical practitioners, who wrote their respective letters, that they had lived in the female role, but had not. Further, Kassandra reported that the practice of purporting to be living in the female role and not is a common practice amongst the transsexual community.

The surgery is superficially impressive. A neo vagina has the cosmetic appearance of a vagina of a natal female. Unfortunately, the neo vagina falls far short in terms of function. The neo vagina will not fool any medical practitioner when examined closely, despite the claims of many plastic surgeons. The procedure doesn't provide any ability for the neo vagina to become moist for intercourse. There are no glands that service the neo vagina to affect this function unique to females. Further, the neo vagina does not begin to address the other sexual organs of the body that foster procreation. Procreation is a critical function of the female sex and part of what distinguishes the female of a species from the male. Further, the people, who undergo the procedure, must apply stents to the cavity as the male body naturally attempts to reject it. If a post-operative, male-to-female transsexual fails to apply stents to the area, the body will close the opening of the cavity, like any wound to the body. If that happens, the person effectively has no apparent genitalia whatsoever.

Raven is one of eight patients Dr. Brassard is scheduled to see this week. As of 2004 Dr. Brassard charges $16,800 for a gender reassignment surgical procedure, not including Goods & Services Tax (GST). The GST amounts to $1,176, bringing the total cost of the procedure to $17,976. Dr. Brassard reportedly requires full payment prior to performing the surgery.

Dr. Brassard is known to the transsexual community as one of the best. There are a number of testimonials on the walls of the Montreal Clinic from former patients of Dr. Brassard. A number of the letters express heart felt appreciation to Dr. Brassard to have undergone the procedure from him. The letters talk glowingly about how well the patient is doing three months, six months and even one year after undergoing the surgical procedure.

Raven has met with Dr. Brassard on at least one occasion prior to today to talk about what he could do for Raven. Brassard refers to this meeting as a consultation and deducts one day of recovery time at the estate, if candidates for the surgery opt to have one. During a consultation Dr. Brassard carefully walks a candidate through the procedure and what he can do and can't do, in addition to answering any questions a candidate may have. In Raven's case Dr. Brassard said he could fabricate a life-like labia minora where Raven's male genitalia now exists.

Prior to the surgery Dr. Brassard's staff gave Raven a fact sheet for post-surgical care. It is very important that the patients follow the post-surgical care instructions to avoid post-surgical complications otherwise Brassard's patients will have to pay more for subsequent visits.

Prior to the surgery Dr. Brassard's staff have carefully cautioned Raven that there is a strong probability that Raven will lose all sexual function and will never know what it feels like to orgasm again. In fact a study conducted in 1999 found that 29% of all males, who reported sexual function as being important and who underwent a gender reassignment surgical procedure, report they now can no longer orgasm. [2] Raven, who is well read in these matters, is undeterred.

Prior to the surgery Dr. Brassard's staff also made Raven aware of the risks and complications surrounding the surgery. One of the most dramatic risks and complications of the surgery is blood clotting, which can cause death. In fact the clinic has had a patient die from post-surgical complications after undergoing a gender reassignment surgical procedure. In May, 2002 or thereabouts, Michelle Renee died days after undergoing the procedure at the Montreal clinic. Dr. Maynard and not Dr. Brassard performed the radical surgery on Renee. The risks also include urinary tract infections and complete loss of sensation of the genital region. Raven, who is aware of the risks, is undeterred.

Prior to the surgery, Dr. Brassard's staff carefully walked Raven through a number of legal documents that he must sign prior to undergoing the procedure. The documents effectively protect Dr. Brassard from any liability, stemming from the surgery, including post-surgical complications and medical malpractice. Raven, who is aware of the risks of having no assurances of post surgical success and the risks of having no recourse to seek legal remedies if the surgery should fail, is still undeterred. Raven reportedly has put his complete faith in Dr. Brassard and his surgical experience.

When the Fifth Estate interviewed Brassard and asked him whether a gender reassignment surgical procedure is in the health interests of his patients, he reported that he doesn't know. Brassard reported that he only knows that his patients say that they 'need' his services by the time they arrive at his private clinic.

Raven is not at the clinic alone. 'Shanin', Raven's partner, is with him. Shanin (not his real name) is also undergoing a gender transition. It was only one year ago that Shanin answered to the name, Sean, and reported having a bisexual history. In fact, when Sean changed his name, he published a long essay, describing the reasons why he elected to change his name and why he continued to answer to the name, Sean, even though he went to work presenting as a woman. [3]

Shanin is also scheduled to undergo a gender reassignment surgical procedure with Dr. Pierre Brassard, however, Shanin's appointment to undergo the procedure is scheduled in September, 2005, months after Raven's surgery. Following the study in 1999, the so-called 'lesbian' couple have a 42% chance that both will be able to relearn how to orgasm after they completely recover from the procedure. Yet, presumably both partners believe the surgery is in their common law interest to undergo.

Fortunately, after the surgery and after a short recovery period, Raven reported that he (now legally 'she') is satisfied with the surgical result and that he "feels better" post-operatively. Brassard kept to his promise to Raven to fabricate a life-like labia minora where male genitalia had existed before. Raven did not report at the time whether he was ever able to orgasm as a post-operative transsexual woman. Further, Raven never reported at the time whether, prior to the surgery, he could effectively orgasm. The introduction of synthetic hormones, and in particular anti-androgens, which Raven reported taking to flat line his free testosterone level, chemically castrates males over a few short months of time. Raven, who in all likelihood was chemically castrated going into the surgery, would likely have been no exception.

After the surgery Raven reported he got a 'decent' job as 'she' and with his future looking bright he treated himself and underwent a facial feminization surgical (FFS) procedure from Dr. Ousterhout. Raven reported the FFS and not necessarily the GRS procedure dramatically changed his life. In fact, Raven attributes the FFS procedure as the final step in completing his gender transition.

As of 2015, ten years later, Raven and Shanin were no longer a romantic couple. The two had by that time parted way. In 2015 Raven was married to a natal female.

This book, however, is not about Raven or Shanin or the thousands of people, who undergo a gender reassignment surgical procedure in North America each year, and 'feel better' after the surgery or whether these individuals ever experience orgasm after radical surgery to their genitalia. This book is also not about Dr. Brassard or the dozens of plastic surgeons like him worldwide, who earn millions of dollars (gross) from medical services geared towards transsexuals without any knowledge of whether a gender reassignment surgical procedure is in the health interests of their patients. This book is also not about the clinical psychiatrists, who approve men and women to undergo irreversible surgical procedures to alter their sexual anatomy, many of whom take their patients at face value and barely have any knowledge of the field.

This book is about the thousands and thousands of men and women, who five years after undergoing a gender reassignment surgical procedure fail to benefit from the procedure, and who respectively watch their life get worse and not better. This book is also about the thousands and thousands of men and women, who after undergoing a gender reassignment surgical procedure, commit suicide each year for reasons that have nothing to do with prejudice. This book is also about the thousands and thousands of men and women, who after undergoing a gender reassignment surgical procedure, have effectively dropped out of society and no longer project any gender expression whatsoever—masculine or feminine.

* * * * *

It is Sunday, July 4, 2004. Toronto, Ontario is hosting the twenty fourth edition of the Gay Pride Parade. The event otherwise celebrates sexual diversity and is widely attended. In fact, the Toronto Gay Pride Festival is the third largest of its kind in the world and draws hundreds of thousands of visitors to the city year after year. This year the event has drawn an estimated 800,000 spectators. Kaitlyn is one of the spectators in attendance at the parade.

Kaitlyn is a post-operative, male-to-female transsexual. Kaitlyn reportedly admitted that attending the parade in the past didn't make him female, but he likes to attend the parade anyways. [4] Kaitlyn reportedly also admitted that undergoing a gender reassignment surgical procedure didn't make him female, nor did undergoing a breast augmentation procedure, nor electrolysis, nor hormone replacement therapy nor anything he tried to physically change about himself.

Prior to his gender transition, Kaitlyn worked for Purolator. In fact Kaitlyn had a successful career at Purolator. Kaitlyn reported that he advanced to a unionized, team leadership position at Purolator, managing a number of specialists at one of the company's largest clients. Kaitlyn got good reviews for his work and realized significant wage increases at Purolator prior to his gender transition. In the autumn of 1995, Kaitlyn announced that he intended to undergo a gender transition and change his identity and appearance to that of a female. Kaitlyn reported that at first Purolator, who had instituted a diversity program, didn't know what to make of it.

Kaitlyn reported that a director in Human Resources was very supportive of Kaitlyn and that the director supported Kaitlyn 100%. However, the company reportedly believed that Kaitlyn's gender transition may have an adverse affect on the client relationship where Kaitlyn worked. Kaitlyn, appreciating the company concerns, voiced a willingness to be transferred to another division of the company—outside the sphere of direct client interaction. The company and Kaitlyn agreed that the transfer would make things 'easier'.

Purolator offered Kaitlyn an internal job. The job was at a reduced salary, which Kaitlyn agreed to. The position was also non-unionized which meant if Kaitlyn wanted to stand up for his rights and dignity, he would have to do so without the support of a union. Kaitlyn began the gender transition at the outset of his new assignment. Unfortunately, things reportedly didn't go well. Kaitlyn's coworkers at Purolator reportedly wanted little to do with him. In fact, Kaitlyn coworkers complained. Kaitlyn's manager reportedly pulled him aside and said the new arrangement wasn't working.

Purolator reportedly gave Kaitlyn two options. Purolator offered Kaitlyn an entry level job at the loading dock, which involved yet another reduction in salary, or a severance package. Kaitlyn reportedly surprised the management at Purolator and opted for the job at the loading dock.

Kaitlyn contacted the Ontario Human Rights Commission (OHRC) in Ottawa, Ontario to complain of how he was treated at Purolator. The OHRC assigned a lawyer to the case. The OHRC engaged Purolator in mediation on Kaitlyn's behalf to resolve the problems Kaitlyn was having with the company.

Purolator's lawyers reportedly made the argument that Kaitlyn's gender transition caused a measurable disruption to operation of the business. When the OHRC questioned the cause of the disruption, Purolator reportedly had little grounds other than citing Kaitlyn as "obviously different'. At the conclusion of the mediation, Purolator created a new position for Kaitlyn—a Drop box Auditor. The position effectively involved attending to drop boxes scattered over Toronto, Ontario, ensuring that they were clean; and that they had ample supplies for customers to send packages. The job was extremely isolated.

Kaitlyn situation slowly improved and over the next two years from 1997 through 1998 he received a number of wage increases. Kaitlyn reported that Purolator earned an excess of a quarter million dollars of additional revenue from the drop boxes that Kaitlyn attended to.

However, Kaitlyn reported that the work was physically exhausting. Kaitlyn, who was over age 40 at the time, had to carry in excess of forty pounds of materials as he walked from drop box to drop box. Kaitlyn reportedly began to develop arthritis from his job as Drop Box Auditor.

Kaitlyn attended the gender identity program at the Centre for Addiction and Mental Health (CAMH) throughout the gender transition. In 1999 the psychiatrists and psychologists at the CAMH had reportedly thought Kaitlyn was a good candidate to undergo a publicly funded, gender reassignment surgical procedure and made the arrangements for Kaitlyn to undergo one. On September 9, 1999, Kaitlyn underwent a publicly funded, gender reassignment surgical procedure. The Ontario Health Insurance Plan (OHIP) funded the procedure.

During the gender transition Kaitlyn underwent hormone therapy to change his physical appearance to that of the opposite sex. He also underwent electrolysis to protract his facial hair. He also underwent a breast augmentation surgical procedure. Kaitlyn reported that he was otherwise a heterosexual male.

In 1999, after undergoing the GRS procedure, Kaitlyn reported that Purolator had began to complain about the quality of his work. Kaitlyn reported that Purolator also criticized him over his physical appearance. Kaitlyn's supervisor cited Kaitlyn for wearing dirty clothing while on the job. Kaitlyn reported that from that point forward Purolator no longer gave him any wage increases.

Unfortunately, things at Purolator reportedly did not improve. Kaitlyn reported being in poor physical health. He also reported that he had difficulty with 'mood swings'. In December, 2000, (or thereabouts), one year after undergoing a gender reassignment surgical procedure, Kaitlyn reported that he suffered from depression. Kaitlyn offered to resign from Purolator in exchange for a severance package. Purolator took Kaitlyn up on the offer. On February 9, 2001 Purolator terminated Kaitlyn's job with the company and gave him a severance package.

After the termination from Purolator, Kaitlyn struggled to find work. In 2005 Kaitlyn, who regularly attends the meetings of a support group at the CAMH, reported to a number of members of the support group that his financial situation had become 'critical' due to his inability to find work. In 2005 Kaitlyn reportedly applied for and began to receive social assistance to support himself financially.

At the support group sessions Kaitlyn reportedly no longer wore gender-appropriate clothing although did prior to undergoing the GRS procedure. Kaitlyn explained that he only wore gender appropriate clothing during the gender transition, which he felt was over, and it was now no longer necessary to do so. At one point Kaitlyn let his facial hair—what was left of it—grow out. A number of members of the transsexual community reported that Kaitlyn indeed looked 'awful'. However, Kaitlyn didn't see himself the way others saw him. In fact, Kaitlyn reported being a 'passable' transsexual, who had the identity of a 'butch dyke'. No other person who lived in Kaitlyn's neighborhood reportedly thought Kaitlyn was 'passable'. In fact Kaitlyn's neighbors reportedly regarded Kaitlyn as a man and transsexual. A number of Kaitlyn's neighbors reported Kaitlyn as a 'guy'. None of the people in Kaitlyn's community regarded Kaitlyn as a woman.

Kaitlyn otherwise reported that he was satisfied with the results of the GRS procedure and that he had no regrets undergoing the radical surgery and if he were ever face with the choice of doing it all over again he would choose to undergo the surgery in a heartbeat.

As of 2015 Kaitlyn's linkedin.com profile includes only one work engagement, that of a janitor, with a duration in excess of one year.

Kaitlyn's behavior of wearing facial hair post-operatively is not unique or unusual. In fact a number of male-to-female transsexuals, who undergo surgical procedures to alter their gender appearance, wear facial hair and no longer project any gender expression whatsoever.

Kassandra, a post-operative, male-to-female transsexual, reported she knew a male-to-female transsexual, who wears facial hair.

The behavior is also not isolated to otherwise heterosexual male-to-female transsexuals. A number of homosexual male-to-female transsexuals also exhibit the same behavior post-operatively. However, in the case of the homosexual male-to-female transsexuals, the root cause of the behavior is the thought of functioning as woman during intimacy and not necessarily the male partner.

In fact, virtually every male-to-female transsexual knows of a post-operative, male-to-female transsexual, who no longer wears gender appropriate clothing or cosmetics, but did during their gender transition; virtually every male-to-female transsexual knows of a post-operative, male-to-female transsexual, who no longer projects any gender expression whatsoever. Further, any member of the transsexual community, who claims otherwise, is either oblivious or deliberately being evasive.

* * * * *

It is Friday, July 30, 2005. The OHHO Gallery physically located at 348 Danforth in Toronto, Ontario is featuring an exhibit of a number of local artists. The exhibit is scheduled to run at the gallery through August 12, 2005. Jennifer Claire Pallister is one of feature artists. Pallister is a male-to-female transsexual, who identifies post-operatively. Pallister took first place in the 2004 Visual Arts category of the Frankly Bob Awards. [5] The Frankly Bob Awards recognize outstanding achievements in Literature and Visual Arts for persons who experience homelessness or long term poverty. The awards are organized by the Neighborhood Innovations Fund and the Silver Dollar. The prize for first place for any category is $300.00. The prize for honorable mention is $50.00 dollars. In 2003 Pallister received an honorable mention for visual arts. The web site that publishes past winners of the Frankly Bob awards no longer publishes the 2004 winners or the fact that Pallister won that year in the visual arts category.

Pallister is hoping to develop a second career in art. Pallister reportedly studied under Theresa Morin or Natalie Wood of the Inspirations Art Studio at 761 Queen Street West in Toronto, Ontario. The Inspirations Art Studio is a CED entrepreneurial initiative operated by

Sistering. Sistering is a non-profit outreach organization, whose mission is to support homeless, under-housed and low income women in the Toronto community. [6]

Jennifer Pallister was born Randy Pallister. Prior to his gender transition, Pallister worked as a software engineer at Symantec Delrina. [7] Symantec Delrina is a world-class software vendor. At Symantec Delrina Pallister worked as a software engineer on printer drivers, user interface controls and on its award winning product, WinFax. Delrina sold millions of licenses of WinFax, embedded with Pallister's code. If Pallister's code didn't work—which happened at times—the company lost in his words, "thousands and thousands" of dollars per day. Pallister reportedly earned a higher-than-average salary, drawing $60,000 per annum at Symantec Delrina. In 1994 the average salary for a software developer, working in Toronto, Ontario was approximately $40,000 per annum.

Pallister reported that he began a gender transition at Symantec Delrina in September, 1994. Pallister reported going to work in women's clothing, including blouses and skirts. He also wore a wig to hide a receding hairline. Pallister reported he felt enormous freedom, when he went to work in women's clothing. In fact, Pallister did not pull anyone in management aside to tell them what he planned to do. He reportedly just went forward with it.

Pallister reported that his coworkers were generally supportive at work. Pallister reported that one of his coworkers complimented him on how good he looked as a woman. In fact, one individual reportedly said he looked "better" as a woman than as a man. However, Pallister's energies were more focused on his gender transition than work. Pallister reported that his productivity at Delrina had measurably waned. In November, 1994, Delrina dismissed Pallister, citing productivity issues. Pallister did not engage Delrina in any grievance for wrongful dismissal or complaint as a result of being terminated. In fact he acknowledged the productivity issues.

Prior to working for Symantec Delrina, Pallister reported that he had an erratic work history, writing software. Pallister cycled between having full time employment and being unemployed for significant stretches of time typically one to two years.

Pallister reported that in 1994 he entered the gender program at the CAMH. At the time the CAMH was known as the Clarke Institute of Psychiatry. Pallister reported that the CAMH referred him to Betty Chan, M.D., an endocrinologist, to undergo hormone therapy. After two months, Pallister stopped seeing Chan, who he didn't like, and sought out the care of general practitioner, Dr. John Goodhew. Pallister reportedly didn't see eye-to-eye with Goodhew either. Pallister reportedly regarded Goodhew as a 'maniacal prick'.

During his gender transition, Pallister attended transsexual support group meetings on a bi-weekly basis at the 519 community centre in Toronto, Ontario. The 519 community centre is located in the heart of Toronto, Ontario's gay district. There he met life partners, Shadmith and Crystal. Shadmith is a male-to-female transsexual; Crystal is a natal female. From meeting the so-called 'lesbian' couple, Pallister identified with the lesbian lifestyle where one partner is a male-to-female transsexual while the other a natal female. Shadmith's working relationship was living proof that a relationship with a natal female was possible.

By 1998 Pallister, who had grown impatient with the psychiatrists and psychologists at the CAMH, reportedly obtained letters from one psychiatrist and one psychologist and privately engaged Dr. Eugene Schrang, M.D. in Leeham, Wisconsin, to undergo a gender reassignment surgical procedure. When Pallister abandoned the CAMH, he effectively also abandoned any chance of receiving public funds towards the procedure through the Ontario Health Insurance Plan (OHIP).

Pallister reportedly underwent the first stage of procedure by Dr. Schrang, M.D. in Leeham, Wisconsin, on June 21, 1998 and the second stage on November 2, 1998, a few months later. Pallister also underwent a procedure to repair a receding hair line from another plastic surgeon. Soon after undergoing the two procedures Pallister legally changed his legal identity and his sex code to that of female.

In total Pallister spent $55,000 on the two procedures. Pallister reported that he (now legally 'she') financed the cosmetic procedures through the sale of a condominium property, which he sold, realizing a capital gain of approximately $50,000 (or thereabouts). Despite the sale of the property and after paying for both procedures Pallister still owed $5,000 on a VISA account.

Despite having a great resume to draw from, Pallister struggled to find a job as 'Jennifer' to replace the position at Delrina. In fact after the surgery, Pallister reportedly failed to ever again find full time employment as a software developer.

After the surgery, Pallister applied for and received long term disability benefits from the Workplace Safety and Insurance Board (WSIB) then known as the Workman's Compensation Board (WCB). Pallister reportedly lived off the disability benefits, which barely covered his rent and groceries. At one time, when Pallister couldn't make ends meet, he sought refuge at a woman's homeless shelter in Toronto, Ontario.

Pallister had a history of mental illness independent of a gender identity disorder. Pallister reported having a borderline personality disorder. Pallister also reported that he suffered from an attention deficit disorder as a child. Further, Pallister reported that he suffered from episodes of depression.

Pallister reported that he was otherwise heterosexual and engaged in sexual relations with women. Pallister reported that in High School he engaged in cross-gender role play, where in one instance he acted out having a vagina while he directed the woman to act out having a penis. Pallister reported that in the case of one woman, who had an Indian background, he not only wanted to have sex with the woman, but he wanted to look like her as well. Pallister characterized the woman as 'exotic'.

Unfortunately, the art show, that ended on August 12, 2005, did not launch Pallister into a second career as an artist. The art show, that ended on August 12, 2005, would be Pallister's last. Pallister reportedly died of complications from alcohol abuse on August 29, 2007 at age 46. Members of the transsexual community reported that Pallister suffered from depression at the time of his death.

* * * * *

It is 6:30 p.m., Saturday, May 22, 1999. The Take a Walk at the Wildside Bed & Breakfast is in full swing. There are no less than eight male cross-dressers in the third floor powder room, preparing for a night on the town cross-dressed. Husband and wife team, Tom Sloan, a.k.a., Roxy Wildside and Patricia Aldridge operate the Bed & Breakfast. The Take a Walk on the Wildside Bed and Breakfast is well-known to largely straight, closeted, cross-dressers across North America and Europe. 'Lori', a close friend of Aldridge, is also at the Bed & Breakfast.

Lori is a post-operative, male-to-female transsexual. Prior to his gender transition Lori was a straight male, who like many of the other men here this evening, changed at the Take a Walk on the Wildside and went out in public cross-dressed.

Prior to his gender transition Lori worked at a government ministry. Lori developed software and also provided network support at the ministry he worked. Lori reported that he was highly regarded professionally. In fact he reported that the senior management team gave him the expectation that he had what it takes to climb the management ladder in government.

Lori reportedly entered the program at the Gender Clinic at the CAMH in 1994 or thereabouts. At the time the CAMH was known as the Clarke Institute of Psychiatry. Lori reported that the CAMH had diagnosed him as having a gender identity disorder. He reported that at the time he believed the diagnosis was correct. In any event, the CAMH under the direction of Psychologist Ray Blanchard assessed all heterosexual males, seeking a gender reassignment surgical procedure, as autogynephilic transsexuals. Ray Blanchard coined the term, autogynephilia, to denote a male's paraphilic tendency to be sexually aroused by the thought or image of himself as a woman. Blanchard, himself, is of the opinion that autogynephilia in isolation does not necessarily disqualify a candidate, seeking a gender reassignment surgical procedure—which Lori sought.

Lori successfully transitioned at work and met all the requirements to undergo a publicly funded GRS procedure. Lori presented as a woman on a daily basis, wearing cosmetics and gender-appropriate clothing, including blouses, skirts and women's shoes. Lori underwent a gender reassignment surgical procedure in 1998 around the time the Mike Harris government stopped publicly funding the procedure, but since Lori was approved for the surgery prior to the cut to the funding, the cost of Lori' surgery was grandfathered.

Lori had a number of physical ailments, including Crohn's disease, which forced him to take time off work. In fact, Lori reported being on long term disability at the ministry a number of times. However, none of the physical ailments Lori reported had any relation to gender identity or his prospects for advancement.

Lori reported that with the gender transition his opportunities for advancement at the ministry evaporated.

Years after the surgery Lori began to question whether he was properly diagnosed by the staff of the CAMH. Lori reported to a number of members of the transsexual community that he does not have a natural gender identity disorder, but rather, an autogynephilic disorder and it was autogynephilia that motivated him to undergo a gender reassignment surgical procedure. In any event Lori never expressed any regret undergoing the radical procedure that amputated his male genitalia.

references

[1] gender pronouns

In the book I utilize male appropriate pronouns to describe what are known as male-to-female transsexuals and female appropriate pronouns to describe what are known as female-to-male transsexuals. The reason is to provide clarity with respect to the person's biological anatomy.

It is legally possible in North America for individuals to change their sex code. The requirements differ from jurisdiction to jurisdiction. It is generally politically correct to respect transsexuals by using pronouns that are consistent with the gender they identify themselves to be. However, a person's legal sex and anatomical sex are not necessarily the same.

Despite the advances in medical science, no technique or procedure exists to change an individual's anatomical sex. Medical science has had some success altering a person's outward gender appearance through the introduction of synthetic hormones. Medical science has also had some success in cosmetically fabricating the anatomy of a penis into what looks like a vagina. Medical science has also had some success in cosmetically fabricating a male penis by way of what are little more than stents and skin grafts. These medical successes, however, are extremely limited.

Further, the medical advances fall far short of satisfying the basic definitions of what is female of a species and what is male.

To illustrate the point in the case of a male-to-female transsexual, when he undergoes a penectomy and orchiectomy, he disables his ability to fertilize, which is a critical function of the male species, but he does not advance his ability to conceive or bear offspring, which are critical functions of the female of the species. In the case of a female-to-male transsexual, undergoing testosterone disables her ability to conceive, however, the attachment of a stent does not advance her ability to fertilize, which is a critical function of the male of the species.

Even the most 'passable' transsexuals have moments when they fail to pass. {Maxine Peterson, formerly Leonard H. Clemensen, M.A., CAMH, 2002]

[2] Rehman J., follow up study on 28 post operative transsexuals

The reported sex and surgery satisfactions of 28 postoperative male-to-female transsexual patients; Archive OF Sexual Behavior; (1999) 28:71-89, Rehman J. Lazer S., Benet A.E., Schaeler L.C., Melman A.
more information
visit the site: Springlink Abstract of the Study
http://www.springerlink.com/content/p005571hmv827611/
visit the site: Springlink Orgasm in Post Operative Transsexuals
http://www.springerlink.com/content/p4347n067550n604/

[3] Shanin's Notice of Name Change

On May 18, 2003 Sean Blatt publishes a publication, notice of name change.
more information
visit the site: Notice of name change
See Appendix A.

[4] gender pronouns

See [1]

[5] Frankly Bob Awards

The Neighborhood Innovations Fund in partnership with the Toronto Dollar organize what are known as the Frankly Bob Awards. The Frankly Bob Awards are an annual event that recognizes outstanding achievements in Literature and Visual Arts for persons who experience homelessness or long term poverty. The Toronto Community Foundation, which is non-profit charitble organization, takes donations for the awards.
more information
visit the site: Frankly Bob Awards
http://www.torontodollar.com/events/2005/franklybob/index.php

[6] Sistering

Sistering is a non-profit outreach organization, whose mission is to support homeless, underhoused and low income women in the Toronto community.

more information
visit the site: Sistering
http://www.sistering.org/

[7] Interview with Jennifer Claire Pallister, dated February 8, 2000

I conducted an interview with Jennifer Claire Pallister on February 8, 2008. The audio material is available via the link below.
Interviews
Pallister Interview, dated Wednesday, February 8, 2000
http://www.transgression.com/assets/download/interview%20Jennifer%20Pallister,%20dated%2020000208.mp3
certified transcript of the Jennifer Pallister Interview, dated October 12, 2005
http://www.transgression.com/assets/downloads/transcript%20interview,%20pallister,%20dated%2020200002 08,%20edited.pdf

CHAPTER 02

Sex & Gender

Transsexualism and the causes underlying transsexual behavior touch upon two core concepts, including Sex and Gender. The relevant definitions of the word, sex, include:

> 1 : either of the two major forms of individuals that occur in many species and that are distinguished respectively as female or male especially on the basis of their reproductive organs and structures.
>
> 3 a : sexually motivated phenomena or behavior.

The first definition of the word, sex, deals with anatomy and how that anatomy plays into procreation. The critical function of an adult male of any species, having two distinct sexes, is the ability to fertilize. The critical function of an adult female of any species, having two distinct sexes, is to conceive and bear offspring. In mammals the adult female brings a fetus to term.

The third relevant definition of the word, sex, deals with the behavior that surrounds procreation. Sex (as a motivated phenomenon) is irrepressible and encoded into any species having two distinct sexes.

The relevant definitions of the word, gender, are broader and deal with traits involving behavior. They include:

> 2 a : sex b : the behavioral, cultural, or psychological traits typically associated with one sex

The definitions of the two words overlap somewhat. One of the definitions of the word, gender, (2 a) is sex. However, it is unclear which context the definition refers to. For the purposes of the book I will presume the definition of the word, gender, (2 a) is defined as the anatomical sex of a member of a species.

The third relevant definition of the word, sex, (3 a) and how it plays into transsexual behavior is in fact the most hotly debated topic inside psychiatric and psychological circles. The third relevant definition of the word, sex, and how it plays into transsexual behavior is also the most despised topic inside the transsexual community. (see Chapter 4, Men Trapped in Men's Bodies). Further, when sexologists, academics, researchers, or members of the media put forward evidence of a (arousal-based) sexual motivation rather than a gender-based motivation, underlying transsexual behavior, a number of members of the transsexual community take steps to intimidate and harass the people, who do (see Chapter 5, the Practice of Fear and Intimidation).

The forces that shape a woman from a man are a series of incredibly complex events that largely occur in the mother's womb. The process of fetal development is the end result of tens of millions of years of natural selection and evolution. Here—in the microscopic world inside the human uterus, timing is everything. Chromosomes provide the simple Xs and Ys of the human genome to determining sex.

A human egg carries a single X chromosome. Whereas the sperm from the father either carries an X chromosome or alternatively a Y chromosome, the statistical proportion of X bearing sperm and Y bearing sperm are significant depending on the environmental conditions of these fields of fertilization. Acidic properties in the female's vagina can selectively favor sperm bearing X Chromosomes over their Y counter parts. This may explain why the human species has more females than males.

When the egg is fertilized, the chromosome in the sperm determines the physical sex of the fetus. But the chromosome pairing (XX or XY) won't always match the fetus' physical appearance.

During the critical first trimester, the embryo's sex-related genes go to work. A gene on the Y chromosome triggers a flood of androgens from the embryo's developing testes. Androgens, such as testosterone, are steroid hormones—chemical messengers that eliminate potential femininity and ensure that the fetus develops into and functions like a "normal" male.

A genetically female embryo's adrenal glands (which produce hormones) also secrete androgen, but in smaller quantities. Instead, the embryo (with the help of the uterus) produces—and is bathed in—loads of estrogen. By the ninth week, the ovaries of a female are neatly tucked into place; by the end of the 12th week, an ultrasound reveals a female. But if too many androgens are floating in the uterus, an XX fetus can end up with male equipment. That's where intersexuality occurs. Again, timing is everything. The process usually works amazingly well. To call this the miracle of procreation is perhaps an understatement. Modern science of the twenty first century pales in comparison to what nature does with routine precision.

The forces of biology and in particular the role of hormones in the child continue well after birth. Although males and females are born into bodies that share a common design— breathing, circulating blood, processing toxins— a proportion of the bodies of the two sexes are different. The sex differentiated organs are called sex organs. The sex organs include the prostate, ovaries, uterus, testes, vagina, penis, etc. The brain has historically been seen to fall under the category as being part of that shared design. But is it?

It is not all about biology, but biology is a driver. Where exactly does biology stop and socialization begin? Or does the combination of biology and socialization play off on each other? To illustrate where biology ends and socialization begins consider the concept of the gender appropriate toy. Do we create toys that we think children of different sexes will like or do children choose the toys of their choice and the toy manufacturers simply respond these changing market demands? As it stands lots of boys ask for dolls. Parents generally do not encourage boys to ask for dolls. In fact parents go out of their way to purchase gender-appropriate toys for their children and apply negative reinforcement to their boys, who ask for dolls.

In a study involving almost 300 children, researchers found that if boys asked for a soldier equipped with battle cannons for their birthday, their parents would buy it for them 70% of the time. If they asked for a Barbie doll or similar gender inappropriate toy, parents would buy the Barbie doll only 40% of the time. Attempting to segment biology and socialization is nearly impossible. However, a few researchers attempted to do just that in pre-birth differences where socialization is completely absent. They discovered that male fetuses are more active and restless than female fetuses. However, after the first year of development after birth, toy preferences become distinct. Boy infants tend to opt for toys, promoting mechanical play while girl infants gravitate toward toys with faces, toys that can be cuddled.

The latest theory insofar as "play" is that it is a practice run at adult life. Through games and role play, the experts say, we learn the art of sizing up the competition, how to win and lose gracefully, which leads directly into how to build friendships. In scientific terms we learn socialization.

In the late 1970s, Robert Goy, a psychologist at the University of Wisconsin, first documented that juvenile male monkeys consistently played much more aggressively than juvenile females. Goy then went on to demonstrate that if you manipulate the testosterone level—raising it in females, cutting it off in males—you reverse the behavior of the sexes. Male monkeys began to act in a more gentle manner and the female monkeys began to act in a rough-and-tumble manner. Research have never tried to recreate studies of this nature on the human population. However, researches see this same type of behavior you would expect in girls had you raised their testosterone level in one type of intersexual, the congenital adrenal hyperplasia (CAH), in which a female's adrenal gland inadvertently boosts testosterone levels naturally. Researchers found that CAH girls, in general, prefer trucks and cars and aggressive play. However, they adapt to more gender appropriate play when in the company of normal girls. Yet, if left to choose, they gravitate towards the rough-and-tumble play we are accustomed to seeing in boys.

Higher testosterone levels also affect competitiveness. In fact, testosterone levels are predictable in certain situations. In a competitive situation testosterone levels rise. Researchers have measured the testosterone levels during chess matches, soccer games, and courtroom battles. The testosterone levels in people stay up when they win; and drop when they lose. People who lose literally deflate. The purpose of testosterone is to get you up for competition, which has implications in the animal kingdom—hunting and gathering food and other resources.

Again we see the affects of testosterone in the two sexes in preschool. Boys tend to hang out in larger competitive and organized groups. They play games that have clear rules to determine winners from losers. Boys like to openly boast their accomplishments after victory. Girls, on the other hand, gather in small groups. They engage in theatrical play where there are neither winners nor losers.

However, even with this research the development of a girl to a woman is a complex process. The results from one person to the next can vary tremendously. It varies even more sharply when you regard the socialization process of an individual, who exhibits transsexual behavior.

"Although a great deal is known about the socialization process of girls as they progress towards women, there is a great deal of variation. Class, race, ability all weigh into the development of a female child into an adult woman. Not all women are raised to be unassertive, dependent and nurturing. We know even less about the socialization process of the trans* boy." Eleanor MacDonald says. MacDonald is a professor in the Political Studies programme at Queens University in Kingston, Ontario.

How does gender-based indoctrination affect our gender identity? What does a female child get out of her childhood to equip her to function as a woman in society?

"Children of both sexes learn that there are these two categories of genders and that they are significant. They learn that there is a defined structure. Children take things they are told as concrete and literal. They see gender in the external, for example, they don't understand the adult male with long hair. None of their male friends in school who are their age have long hair, so that presents a violation to what they are taught. This indoctrination constitutes a key part of how you come to understand your gender and your gender identity." Krista Scott-Dixon says. Scott-Dixon is a Ph.D. graduate of the Women's Studies programme at York University. There she concentrated on feminist theorist, feminist epistemology, methodology.

Parental influence is the gatekeeper to the formation of a child's gender identity. Parents ultimately choose what they expose their child to.

"Parents undergo a type of anxiety towards parenting. They are scared of introducing gender inappropriate clothes and toys for fear that that may have damaging repercussions later in life. For example, I gave a friend's child a pink t-shirt with the logo of the sex pistols. The parents were somewhat livid. They didn't want their child, who was a boy, to have a pink t-shirt for fear he would turn out to be gay."

However, what happens when the child isn't listening or if he is, what if he is listening to the wrong message? What if the child is receptive to different signals—signals meant for children of the opposite sex?

"If you are growing up and you are male embodied, you are going to internalize an awful lot of what is being told to girls. You may get a different set of messages. On the one hand [there is] a lot of pressure to grow up and become like a man. [...] on the other hand you are probably also internalizing some of the messages that are being given to your sister or your female cousin or the girl next door about how you are supposed to behave. I don't know if we can say that children growing up who are transsexual are being raised as either male nor female in terms of their social condition. There are probably more interesting combinations than that. It is hard to study because we don't know until well after the fact. However, when you read people's autobiographies about what the experiences were like, it is pretty clear that they were not receiving the same information the same way." MacDonald says.

Still there needs to be a mechanism within the child that instructs to child to switch the learning focus from the people in your defined gender to people, who you are aware as being outside your defined gender. That is where the brain enters.

Not much is known about how the brain plays into the formation of gender identity. In 1997, Dutch researchers discovered that the brain is a sex differentiated organ. In particular the researchers discovered that one region in particular which they label the BSTc as being highly differentiated by sex. The BSTc plays an essential role in masculine sexual behavior, as shown in studies on rats.

In this tiny region of the brain inside the hypothalamus, the neuron density of a "normal" heterosexual male is 44% larger than in a "normal" heterosexual female (2.49 cubic mm compared to 1.73 cubic mm). However, in the same study, the neuron density of a "normal" heterosexual male is 52% larger than in a male-to-female transsexual (2.49 cubic mm compared to 1.64 cubic mm). More importantly, the neuron density of the heterosexual females and the male-to-female transsexual were nearly identical (or within a 5% variance).

The researchers found no relationship between BSTc size and the sexual orientation of transsexuals, that is, whether they were male-oriented, female-oriented, or both. Furthermore, the size of the BSTc of heterosexual men and homosexual men did not differ, which reinforced the idea that the BSTc size is independent of sexual orientation.

The study includes six samples of male-to-female brains. All of the male-to-female transsexuals had undergone some regiment of estrogen therapy, however, researchers discounted estrogen as a cause for the volume of neuron density insidethe BSTc.

The implication of the research is that there may be a physical rational for why people feel a disquiet towards their birth sex. If true, the BSTc is in effect telling the individual that he/she belongs with them over there. However, the standard deviations of the density are significantly large. The study identified a woman, who otherwise had the BSTc density of a normal heterosexual male, but reported no gender discomfort.

CHAPTER 03

Decoupling Sex & Gender

The evolution of care towards transsexuals where gender reassignment surgical (GRS) procedures are the prescribed treatment follows three successive waves. The waves are marked by the prevailing circumstances at the time.

The first wave of care towards transsexuals in North America where GRS procedures are the prescribed treatment is characterized by 'adoption and centralization'. The wave began in 1965 when Reid Erickson, a wealthy female-to-male transsexual, established the Erickson Educational Foundation. The Erickson Education Foundation sponsored a number of studies into transsexualism. That same year, Johns Hopkins School of Medicine established the Gender Identity Clinic. The clinic evaluated and approved people, who exhibited gender variant behavior, to undergo a gender reassignment surgical procedure at the Johns Hopkins Hospital. The staff of the clinic included New Zealand medical practitioner, John Money, Howard Jones and Milton Edgerton. Phillip Wilson became the first patient in North America to undergo a GRS procedure. After the surgery, Wilson changed his name to Phyllis and became known as Phyllis Wilson.

In 1966 the New York Times published the surgical reassignment of a patient through the Johns Hopkins clinic as a success. [3]

The accepted understanding of transsexual behavior is largely the result of the ideas of John Money. In fact, John Money coined the term 'transsexual' to describe an individual having a persistent disquiet with their assigned gender and seek surgical procedures to alter their sexual anatomy. In addition to the differentiation of the sexual anatomy Money believed there were other 'derivative differences' between the sexes. Money identified the concept of a gender role that functions independently of genitalia and non erotic activities.

Over the next two years, a number of hospitals followed Johns Hopkins School of Medicine and opened their own Gender Identity clinics and began to offer reassignment surgery, including Stanford, UCLA, Northwestern, and the University of Minnesota. The staff at the clinic at Stanford included psychiatrist Norman Fisk and plastic surgeon Donald Laub.

The staff at UCLA included Dr. Robert Stoller. The Stanford Clinic and in particular Donald Laub, became renowned for innovation. The clinic developed the recto-sigmoid colovaginoplasty technique , a technique for male-to-female transsexuals, as well as metoidioplasty, a technique for female-to-male transsexuals.

However, the clinic at Johns Hopkins University was recognized as the most prominent. The first surgeons in the U.S. to sign up to perform the surgery learned the techniques from drawings provided by the clinic at Johns Hopkins University, including Dr. Stanley Biber. Despite the prominence of the clinic, the hospital at the Johns Hopkins School of Medicine performed less than an average of ten sex-change operations per year. [3]

The clinics effectively dissuaded general practitioners from attempting to treat patients, exhibiting gender variant behavior, and asked practitioners to refer these patients to these clinics. Care for transsexuals was centralized and carefully administered. Transsexual patients had little control over the course of their treatment. Harry Benjamin and Stanley Biber were two of the few medical practitioners, who operated outside clinics.

In 1969 the Erickson Educational Foundation (EEF) began sponsoring a series of International Symposia on Gender Identity (ISGI). The first was held in London. The EEF scheduled events in Denmark, Yugoslavia, Stanford, Norfolk, and San Diego.

The second wave in the treatment of transsexualism where GRS procedures are the prescribed treatment is characterized by 'retraction'. In 1975 Paul McHugh replaced Joel Elkes as director of psychiatry of Johns Hopkins School of Medicine. At that time McHugh was suspicious of the practice of surgical reassignment to treat people, who exhibit cross-gender variant behavior. McHugh was reportedly not swayed by Money's apparent successes with the John/Joan study, which was in full swing at the time. McHugh suspected that gender reassignment surgery did not ameliorate the mental illness underlying cross-gender variant behavior but rather corroborated with the mental illness. McHugh was also suspicious of the effectiveness of surgical intervention of intersexed infants at birth.

McHugh turned to William Reiner, a residential psychiatrist at Johns Hopkins, to systematically study the individuals, who underwent a gender reassignment surgical procedure through the Gender Identity Clinic at Johns Hopkins University, to determine just how sexually integrated the subjects functioned in society post-operatively.

In 1977 Reiner published a report of his findings. Reiner's report found that the men, who claimed to be female trapped in a male's body and underwent a GRS procedure, were not the well adjusted women in society the staff had predicted they would become. Reiner also found that transsexuals, who underwent the surgical procedure, did not witness their life improve. Rather, Reiner found that the people, who underwent the procedure through Johns Hopkins, continue to suffer from the same emotional problems they had prior to the surgery.

In October, 1979, McHugh directed Johns Hopkins—the first hospital to perform gender reassignment surgical procedures in the United States in 1965 —to discontinue the surgery. McHugh reported that a number of the plastic surgeons, who the University Hospital hired to perform the radical surgery, were 'relieved'.

In looking at the Reiner and Meyer studies, McHugh believed that human sexual identity is constructed by the chromosomes and the embryogenesis humans undergo and not how people are socialized. McHugh believed that male hormones sexualize the human brain. McHugh conclusions also rely on Ray Blanchard work on the concept of autogynephilia. McHugh found sexual dysphoria—a sense of disquiet about one's sexual role—to naturally occur amongst males, who are raised as females in an effort to correct an infantile genital problem, like David Reimer.

McHugh recognized that "transgender" activists—who were allied with homosexual advocates, including Lynn Conway—continued to argue that they are entitled to whatever surgery they want, and that gender dysphoria represents an accurate concept of their sexual identity. However, McHugh also recognized that "transgender" activists abhor the concept of autogynephilia or any suggestion that their interests towards surgical intervention relates to an autoerotic response towards the consideration of undergoing the surgical procedure. McHugh effectively dismissed their objections to autogynephilia due to the fact that these critics, including Conway, had offered little evidence to refute the diagnosis.

McHugh cites John Money's John/Joan study as evidence of the damage of sexual reassignment especially in males (see below). McHugh stated that collaborating with what he referred to as "madness" damaged Psychiatry's professional credibility and wasted valuable scientific and technical resources.

That same year Janice Raymond published the book, Transsexual Empire, the Making of the She-Male. In her book Raymond condemns male-to-female transsexuals for stealing the essence of femaleness. In her book Raymond alleges that male-to-female gender transitions serve to only further oppress women. (See <u>Chapter 6, Feminism & the Feminist Treatment of Transsexualism</u>).

A number of other university hospitals and gender identity clinics followed Johns Hopkins and discontinued offering GRS procedures as a treatment of transsexualism. Other clinics, including Stanford, detached themselves from the university hospitals that established them and continued to offer their services as non-profit clinics.

The Gender Identity Clinic (GIC) of the Centre for Addiction and Mental Health (CAMH) (formerly known as the Clarke Institute of Psychiatry) is an example of a first wave provider. Unlike the Gender Identity Clinic at Johns Hopkins University Hospital, which shut its doors in 1979 to transsexuals seeking surgery, the CAMH continued to offer GRS procedures to qualified candidates. In fact, under the universal health care scheme, the Ontario Health Insurance Plan (OHIP) and the CAMH, not only offered GRS procedures, but also it publicly funded GRS procedures to all of the candidates who qualified. However, it, too, was forced to close its doors to candidates seeking a publicly funded procedure, when on October 1, 1998 the Mike Harris government cut publicly funded GRS procedures from the list of available treatments in Ontario.

By October 1, 1998, the Clarke had performed approximately 1,200 assessments, averaging 40 new case files per year. At any one time prior to October 1, 1998, the CAMH had 250 to 300 active case files. Ultimately, many of these people sought treatment elsewhere. The number of new case files rose over the clinic's thirty-year-history and up to 1998 the clinic was accepting approximately 60 new candidates per year or thereabouts.

Of all the individuals—the clinic has assessed—the vast majority (70%) did not proceed with a gender transition. Rather, the CAMH advanced only a handful of cases—and only the extreme cases; the cases exhibiting dogged determination. Prior to October 1, 1998, the CAMH approved an average number of 9 transsexual candidates for a publicly funded GRS procedure. In fact the number reportedly ranged from as low as 7 to as high as 12. The CAMH's high attrition rate is directly related to the clinic's policy with respect to patient care. In all cases the CAMH instituted a strict Standards of Care (SOC) document that the doctors at the clinic developed independently of the Harry Benjamin International Gender Dysphoria Association (HBIGDA). The two SOC documents were similar, but not exact. The CAMH's SOC effectively dissuades people seeking to alter their sexual anatomy from doing so.

"Prior to October 1, 1998, the Harry Benjamin Standards of Care didn't specify what the real life experience should consist of. Now it does. The clinic's standards included some form of full time employment; or full time student or full time charitable or any combination of those three for a minimum of two years." Petersen says.

Dr. Ray Blanchard is largely credited for the CAMH's conservative approach to dispensing care to transsexual patients, seeking a GRS procedure. Blanchard is also the frequent target of hatred and ridicule amongst transsexual men and women everywhere in North America. Yet, despite the hatred and ridicule, Petersen remains one of Blanchard's most ardent supporters

"I have worked for Dr. Blanchard for nineteen years. I am extremely supportive of him. He is quite aware of the intensity of feelings that transsexuals experience and he is quite sympathetic to their cause. He is trying to sort it all out scientifically, so, other people within the medical community can take his work and cement their own understanding of transsexualism and perhaps further his work. He does not expect that his work is infallible. He knows it is subject to scrutiny. He knows that it is subject to criticism." Petersen says.

In fact Petersen snapped the license-free images of her beloved colleague for his profile on the Wikipedia web site at the URL, wikipedia.org. The images portray Blanchard as a distinguished man, unfazed by his critics.

In any event now the two SOC documents of the CAMH and the HBIGDA coincide.

Prior to October 1, 1998 when the Mike Harris Conservative Provincial government legislatively stopped the CAMH from dispensing GRS procedures, the clinic sent their candidates to one of two English surgeons: Dr.s Dalrymple, and Royle. The CAMH ultimately allowed the patient to decide which surgeon he underwent the surgery from.

"The reason we worked through them was that they were familiar with government funded, health plans. They did not require financing up front. They were willing to wait to be paid. They trusted our health plan. In the end they knew they would be paid." Petersen says.

In fact, Dr. Brassard, Montreal, Quebec, requires full payment up front prior to performing a GRS procedure at his private clinic.

In addition to screening patients for surgery, the CAMH also plays a role in the candidate's integration back into the community. If the candidate had troubles at work, the CAMH was there to extend education and consulting services. However, Maxine Petersen is quick to point out that their clinic's services in this area are becoming less relevant.

"I would say that the loss of one's job was far more the case ten years ago. Now however, I would say it's far more likely that corporations retain employees who transition on the job.

We are seeing that the staff in Human Resource are far more knowledgeable and sensitive to this than they were years ago. They are largely well prepared to handle a situation like this if one occurs at their corporation. They are not treating this as a big deal anymore. That is particularly true for large corporations and government institutions. Much more so than small corporations. The smaller companies are lagging behind in that regard." Petersen says.

The CAMH was not entirely successful in helping their patients retain their positions and avert significant financial losses. The CAMH observed that the candidates, who reported being straight, had the most difficulties.

"It was the gynephiles." Petersen says referring to the type of transsexuals—the otherwise straight males—who had persistent life issues well after surgery.

The CAMH also couldn't cope with the scenario where the patient did not blossom into the well adjusted woman or man that they aspired to be. A number of their patients, who left their jobs with the intention of reintegrating as a member of the opposite sex, experienced enormous hardship.

"In this case these individuals generally left a job, thinking it would be easy to get one while presenting themselves as a female. In fact, it wasn't." Petersen says.

The CAMH found that post-surgical regret, broken down by sexual orientation to be the most troubling statistic. In a research study by the CAMH in 1989, the group found that the incidence of regret was 29% in female-oriented male-to-female transsexual while at the same time 0% in reportedly male-oriented male-to-female transsexuals.

The loss of sensation and particularly the loss of one's ability to orgasm was not a significant factor in post-surgical regret. Petersen says that 85% to 90% of all post-operative transsexuals report that they are able to achieve orgasm.

"If an individual has an emotional commitment to their spouse prior to transition and if the spouse abandons the marriage and if the individual continues to be in love with the spouse that can be another factor in post-operative regret." Petersen says.

The CAMH, however, has no information on the vast majority of its candidates (the 70%), who fail to advance.

"We have no statistics regarding the people who dropped out of the program and elected to proceed with their surgery in a private clinic. I am not sure if that information exists anywhere. The problem is once someone drops out of the program, we lose touch with them." Petersen says.

The figure is easy to estimate. The CAMH reports 60 new case files per year or thereabouts. 30% of new cases files advance, leaving 18. A number, who never undergo a GRS procedure, are partially autogynephilic. The incidence of partial autogynephilia is approximately 15%, leaving 15. An average of six candidates drop out of the program and advance to undergo a GRS procedure per year.

Now compare that to the entire population of male-to-female transsexuals, who undergo GRS procedure per year and who reside in Ontario. Prior to October 1, 1998 Brassard & Maynard saw the vast majority of these candidates (70%) in addition to the candidates, who reportedly never transition under the watch of the CAMH, but who sometime later undergo the surgery. The other 30% either go to surgeons in the US or to Thailand to evade the rigors of the Real Life Test. Brassard & Maynard jointly perform approximately 400 GRS procedures per year. 1 in 8 (12%) are from Ontario, leaving approximately 48 patients per year, who reside in Ontario. If 70% elect to undergo the procedure at the Montreal clinic, a total of 69 transsexuals, who reside in Ontario, elect to undergo the procedure outside the sphere of the CAMH each year.

Given the gap (69 versus 6), the vast majority of male-to-female transsexuals, who reside in Ontario and who undergo a GRS procedure, in fact elect to circumvent the CAMH (approximately 10 to 1).

As the centralized gender identity clinics put greater and greater restrictions and impediments towards access to hormones and GRS procedures, independent private physicians began offering primary care alternatives towards transsexuals. These independent, rouge practitioners operated completely outside the control of the centralized gender identity clinics that had the mandate to provide care towards transsexuals. However, these general practitioners filled a gap left by the centralized gender identity clinics, who otherwise alienated transsexuals with greater restrictions towards access to hormones and GRS procedures.

General practitioner, Dr. John Goodhew of Toronto, Ontario, who began offering hormone replace therapy to transsexual patients outside the gender clinic of the CAMH in the early 1990s is an example of a second wave primary care provider. Unlike the centralized gender identity clinics, which require their candidates to live in the opposite gender role for one year prior to the introduction of HRT, if the patient has no major ancillary mental disorder, Goodhew starts his patients on HRT after one initial screening. The CAMH, which witnessed its market share erode by the independent private practitioners, attempted to reach out to these independent private practitioners, pleading with them to cease all care to their transsexual patients and instead refer their transsexual patients to them. This strategy ultimately failed and the independent private practitioners stayed loyal to their patients, who otherwise begged them for treatment alternatives to the centralized gender identity clinics, which they felt had betrayed them. In the process these independent private physicians appeased the market demand of the transsexual community for care with less restrictions, not more.

The third wave in the treatment of transsexualism where GRS procedures are the prescribed treatment is characterized by 'decentralized propagation'. Despite the rigorous changes insofar as qualifying candidates for surgery, despite the psychiatric community's increasing suspicions towards GRS procedures as an effective treatment towards so-called gender dysphoria, the number of practitioners who entered the field increased, keeping pace with the market demand. Despite the rigorous changes insofar as qualifying candidates for surgery, despite the psychiatric community's increasing suspicions towards GRS procedures as an effective treatment towards so-called gender dysphoria, the number of patients, who underwent GRS procedures, also increased due largely to the entry of private practitioners and community health centres, who otherwise operate independently from the gender identity clinics. The Sherbourne Health Centre in Toronto, Ontario, which offers a transsexual health program, is one example of a community health centre geared to the needs of transsexuals, which operates independently from a recognized gender identity clinic.

In 2003, the Ontario Ministry of Health gave the Sherbourne Health Centre, located in Toronto, Ontario, a mandate to provide primary care health services to the city's LGBT community. Since receiving the mandate, the Sherbourne Health Centre provides health services to transsexuals across Ontario. After the first year in operation, the clinic took on over 120 new male-to-female transsexual patients, seeking primary care. In comparison the established Gender Identity Clinic of the CAMH on average takes sixty new patients per year. The proportion of male-to-female transsexual to the entire population of case loads is 8%, however, the proportion of transsexuals at the Sherbourne is growing and the proportion of the transsexuals is becoming more important politically. In fact more and more transsexuals in Ontario are opting for care at the Sherbourne Health Centre in lieu of the established Gender

Identity Clinic at the CAMH. The Sherbourne Health Centre is taking twice as many new patients per year than the CAMH.

Although no one I interviewed at the Sherbourne Health Centre would comment on the natural competition that exists between the two heath care providers, a competitive rivalry clearly exists. The Sherbourne Health Centre —which received its mandate in 2003—is part of the third wave of care towards transsexual patents, while the GIC at the CAMH—which received its mandate in 1967 —is part of the first wave of care.

The differences in the approach to care for transsexuals between the two rivals could not be more evident. The established GIC at the CAMH requires its candidates for GRS procedures to undergo a rigorous assessment. The assessment at the clinic includes interviews with two staff psychiatrist and three staff psychologists. After the candidate's assessment is complete the clinic issues the candidate a report, recommending a course of care. The Sherbourne Health Centre, on the other hand, does not require patients to undergo an assessment. The primary care practitioners largely take their patients at face value. Anyone who demonstrates a persistent desire to undergo hormone replacement therapy for over three months and does not have an overt ancillary mental disorder is accommodated.

Leslie Shanks, M.D., the Medical Director, of the Sherbourne Health Centre is at the heart of the health care provider's market-driven approach towards transsexual care.

"One of the reasons we are seeing a number of trans* women coming to our clinic is because we try to be trans* friendly; we try to be trans* sensitive. A number of the people working here are trans* themselves. A number of the staff are queer." Rupert Raj says. Rupert Raj, who is himself a female-to-male transsexual, is a psychotherapist at the Centre.

Shanks is also the driving force behind the centre's flexibility in its cares towards transsexuals, given no one hormone replacement therapy protocol is appropriate for all transsexuals. Shanks reportedly researched the various hormone replacement therapy protocols, including the Harry Benjamin protocol, the Callum Lordes protocol and the Anne Lawrence protocol for the male-to-female transsexuals and made the information available to the primary care physicians.

"85% of the trans* women, who have come through the door, have been on the Internet and have talked to people and have certain expectations of what their treatments are going to be. Whether I can accommodate those expectations or whether I can't [depends]. There will always be several conversations with these patients and what they are thinking as far as a delivery route. If somebody came through my door and said I want to take 10 mg of conjugated estrogens (which is high but is also consistent with Anne Lawrence 's protocol), [I] would talk about that and [ask], ' why do you want to be on that much?' and explore exactly where [the individual] is coming from. If they understand the risks and lack of benefit, I [still] wouldn't put them on that much. I would talk them down a bit." Dr. Keith Loukes says. Loukes, who is fresh out of residency from the University of Toronto Medical School, is a primary care physician at the Sherbourne Health Centre.

The Sherbourne Health Centre clearly fills a gap in the market space, stemming from a general mistrust in the transsexual community towards the GIC due to its reputation of being insensitive and uncaring towards transsexuals. In fact the GIC's approach towards transsexualism—that being to discourage its patients from gender transitions—causes this mistrust.

"This is the most rewarding part of my practice. A number of my trans* patients have come from backgrounds where they experience stigmatization. A number have even had to deal with trans phobia from within the medical community. When trans* patients come to my office, asking for this and that, they quickly learn that that is actually here. I am here to relay back to them that I support that and I am here to help. A number of my patients have left crying; they are so happy", Keith Loukes says.

The Sherbourne Health Centre's positive atmosphere also has tertiary effects on the practitioners. Many report having enormous satisfaction, stemming from the patient interaction.

"I had one patient who was emotional and distraught. She had no idea how to transition. She had difficulty with her family. She didn't have any social support. She didn't even know where to shop. She came in dressed as a woman at her last appointment. I didn't recognize her. I watched her leave happy and excited about spending the day as a woman. She looked fantastic. To me that was like giving birth." Loukes says.

The Sherbourne Health Centre, however, does not have a staff of psychiatrists, who can offer referral letters. The Centre simply has not advanced so far to have the same clout in this area as compared to the CAMH. Ultimately, the Centre refers its patients to the psychiatrists of the CAMH for the final say as to whether the patient qualifies as a candidate to undergo a GRS procedure. Despite the in roads in the marketplace these third wave providers have no alternative but to involve the first wave providers insofar as access to GRS procedures.

* * * * *

Prior to 1966 John Money was a man with a theory about the fluid nature of gender and that gender was mutable at an early stage in a child's development. It was not until 1966 that Money would fall into the landmark study, which he hoped would prove his theories. The case was simply called the John/Joan study.

The John/Joan study innocently began in 1966 with what seemed like a routine trip to a hospital in Winnipeg, Manitoba. Ron and Janet Reimer had planned to have their twin boys, Bruce and Brian, circumcised—a common practice in North America at the time. Unfortunately the procedure on Bruce went terribly wrong and a defect in the machinery caused a burn to Bruce's penis so severe the hospital doctors could not repair the damage. Given reconstructive surgery was not available at the time, the doctors could not offer the Reimer's any options.

The Reimer's watched a television show, months after the accident that featured John Money speaking about gender formation. The Reimers, who were distraught, regarded Money's words as a beacon of light.

By 1966 Money, who was 45-years-old, was a man of his time. Money, who despite being raised in a conservative religious family, was an integral part of the Sixties counter culture and sexual revolution—a revolution that included scientific intrigue; US advances towards landing a man on the moon and back; science fiction film, 2001—A Space Odyssey; and television show, Star Trek.

Money was schooled as a medical practitioner in psychology as well as psychiatry. Money saw himself as a "missionary of sex". Money advocated subjects that were otherwise taboo, including open marriages and bisexual group sex. In fact Money's public statements appear to endorse otherwise abhorrent behavior, including incest and pedophilia. For example, Money believed there were two distinct types of pedophilia, including affectional pedophilia and sadistic pedophilia. Money articulated that affectional pedophilia is about love and not necessarily sex. Money writes:

> If I were to see the case of a boy aged ten or eleven who's intensely erotically attracted toward a man in his twenties or thirties, if the relationship is totally mutual, and the bonding is genuinely totally mutual...then I would not call it pathological in any way.

The Reimers contacted Money and Money wrote back enthusiastically. After listening to the story, Money assured the Reimers that Bruce could be successfully raised as a girl and Money offered to oversee the entire process. The Reimers agreed to move forward with Money's proposition and the twin boys—one the subject, the other the control—would serve as an experiment to prove his theories that gender was fluid and mutable provided the socialization process began at an early enough stage in a child's development. By 1966 Money already had experience in assigning intersexed child at birth, but never a sexually developed child. Money's experiment pitted nurture over nature as pivotal towards gender identity. In fact, Bruce's life and overall happiness would very much depend on Money's radical ideas.

On July 3, 1967 physicians from Johns Hopkins University surgically castrated Bruce and created a small cavity from the remaining skinfold. They called the cavity a "cosmetic vagina cleft." Money did not attempt to have the plastic surgeons fashion Bruce with a neo vagina. Money had planned to have Bruce undergo that procedure later in his development. Money instructed the Reimers to change Bruce's name to Brenda and to never divulge the fact that Brenda was ever not female. At first the Reimers complied.

Early in the socialization process the experiment began to encounter problems. Janet Reimer reported that when she tried to put Brenda into a dress, Brenda objected to the dress and began to rip at it. Brenda was two-years-old at the time. Reimer also reported that Brenda preferred to play with Brian's toys. When Reimer bought toys for Brenda, Brian Reimer reported that his apparent sister used the toys inappropriately. He reported that she would tie people up with a skip rope and not skip rope with it. Reimer also reported that Brenda was constantly the victim of teasing at school by the other children. Brian reported that his apparent sister was the object of teasing on a daily basis and that he characterized the teasing as cruel.

Despite the problems, in December, 1972, Money advertised the John/Joan study as a success at a meeting of the American Association for the Advancement of Science. The event coincided with the release of a book that Money co-authored with colleague Dr. Anke Ehrhardt entitled, Man & Woman, Boy & Girl. The book was the first account of the extraordinary John/Joan study. In Time Magazine Money reported that "this dramatic case provides strong support for the contention of women's liberationists: that conventional patterns of masculine and feminine behavior can be altered. It also casts doubt on the theory that major sexual differences, psychological as well as anatomical, are immutably set by the genes at conception."

The New York Times Book Review hailed Man & Woman, Boy & Girl as "the most important volume in the social sciences to appear since the Kinsey reports" and praised Money for producing "real answers to that ancient question: Is it heredity or environment?"

Despite the incredible story, some of Money's peers were skeptical. Dr. Milton (Mickey) Diamond, a biology professor at the University of Hawaii, persisted to research how the sexual nervous system is organized before birth. He believed in contrast to Money that neither intersexes nor normal, sexually developed children are born psychologically undifferentiated. He was otherwise alarmed that researchers like Money were expanding the practice of infant sex reassignment. Without access to the subjects of the study, Diamond could not dispute Money's claims nor Money's credibility.

Money reportedly planned to see the twins one time per year to "guard against psychological hazards" associated with growing up as a sex-reassigned child. Money saw both Brian and Brenda separately and often together. Unfortunately, the trips added to the questions on the part of the children. The children felt singled out, but didn't know the reason for the attention. Money also showed sexually explicit material to the children to explain the differences between men and women and reinforce sexual identity. Money's sessions with the twins reportedly began to degenerate to horrifying encounters that deeply traumatized the two children. He instructed the twins to engage in what he called "sexual rehearsal play." Sexual rehearsal play involved directing Brenda to stand on all fours on his office couch and for Brian to come up from behind and place his crotch on Brenda's buttocks.

Finally at age nine, Money began to question Brenda about 'her' sexual preferences. Brenda said to Money that 'she' preferred to be with a girl. Money reported Brenda's explicit sexual orientation to Ron and Janet Reimer. They reportedly didn't think being homosexual was necessarily a bad thing.

In 1975, Money published a follow up report, demonstrating the continued success of the experiment. In his report Money neglects to report Brenda's explicit sexual orientation towards females. Money writes:

> No one [outside the family] knows [that she was born a boy]. Nor would they ever conjecture. Her behavior is so normally that of an active little girl, and so clearly different by contrast from the boyish ways of her twin brother, that it offers nothing to stimulate one's conjectures."

In addition to the report to the medical community, in 1975, Money published yet another account of Brenda's successful metamorphosis. In the publication Money addresses a broad audience. Money reported that Brenda's sex reassignment as "dramatic proof that the gender-identity option is open at birth for normal infants."

By the time Brenda reached 'her' teenage years 'she' had attempted suicide at least once. Brenda refused to undergo any additional procedures, but agreed to take an Estrogen supplement, which promoted breast development. The Reimers eventually drifted away from Money and began to seek the support of Dr. Keith Sigmundson, a local psychiatrist in Winnipeg, Manitoba. Sigmundson saw nothing feminine about Brenda, but did not deviate from Money's course of treatment.

Money became keen to fully assign Brenda and have 'her' undergo a GRS procedure. During a visit to Baltimore, Maryland, Money introduced Brenda to a male-to-female transsexual to demonstrate that the surgery Money had planned for Brenda was espoused by an adult. Money didn't account for the fact that Brenda might be shocked to see a transsexual woman, nor appreciate how the male-to-female transsexual benefitted from a GRS procedure. Brenda ran off.

In August, 1979, Brenda, who had been taking estrogens for two years, began to enter male puberty. The synthetic estrogens were now in competition with 'her' natural endocrine system. In effect, despite the absence of testicles, her voice began to break and distinct male-like physical features began to appear. More and more she began to look like the transsexual, 'she' met while at Baltimore, Maryland, whom 'she' feared.

The Reimers transferred Brenda to a technical school where things got worse. Brenda was completely rejected by the other children. When the Reimer's took Brenda to a hospital for examination, a doctor asked 'her' whether she wanted to be a woman. Brenda said emphatically that 'she' did not. During that one visit, Ron reportedly picked up Brenda, took 'her' to get an ice cream cone and told Brenda the truth about 'her' background.

Brenda opted to change 'her' name to David within five weeks of learning the truth about 'herself', effectively rejecting the socialization as Brenda, and opted to undergo a phalloplasty procedure. A phalloplasty creates an artificial penis from skin grafts from the forearm. After undergoing the procedure and after a lengthy healing process, David had a functional penis with limited sensation. David hid in the family basement for a period then emerged as one of the guys. He never looked back.

In 1982 David attempted suicide by taking an overdose of antidepressants, however, Janet intervened and had him taken to the hospital to pump out his stomach. One week after the incident David attempted suicide again. Brian saved his life after the second attempt. David was otherwise reclusive.

David Reimer went on to meet Jane, a mother of three children, and at age 25, he married Jane and adopted her children. He got a job at a factory. Dr. Milton Diamond, who did not get any information from Money regarding the John/Joan study, ran his own advertisements in the American Psychiatric Society Journal. The copy of the ads read:

Will whoever is treating the twins, please report.

Initially, Sigmundson ignored Diamonds pleas. Sigmundson reported that he feared retaliation from Money if he did so. Diamond did, however, catch the attention of one of the doctors, who knew Sigmundson had treated Brenda and gave Diamond his contact information. Diamond reported that the first words Sigmundson said after Diamond contacted him were "'I was wondering how long it would take for you to get here." Diamond, Sigmundson and Ron Reimer collaborated on a follow up study into the famous case study. David Reimer, himself, refused to participate. The ramifications of the reportedly successful reassignment were immense. Plastic surgeons performed thousands of gender reassignment surgical procedures on the advice of psychologists and psychiatrist, who relied on John Money and the empirical proof behind the famous case. In fact, plastic surgeons worldwide performed an average of five GRS procedures per day. Diamond was reportedly determine that when he delivered the truth with what became of the twins of infamous case that he would effectively throw the industry on its ear and that responsible psychologists and psychiatrists everywhere would err on the side of caution. In fact by that time Johns Hopkins School of Medicine had already discontinued the surgery as a treatment for so-called gender dysphoria.

When Diamond and Sigmundson looked for a publisher, journal after journal turned them down due to the controversy that they believed would inevitably ensue. The American Medical Association eventually agreed to publish their findings in the March, 1997 edition of the Archive of Adolescent and Pediatric Medicine.

After the article began to circulate, the practitioners braced for the fall out. A number of critics of the article attempted to dismiss it on the grounds that Diamond is simply using David Reimer's history to embarrass a scientific rival. However, Dr. Melvin Grumbach, a pediatric endocrinologist, published a more measured response. Grumbach believed that embryogenesis does have a measured effect. Grumbach's position is that that sex reassignment should only be done as a last resort and only when all other treatment options have been exhausted.

After David's story came to light in the media, John Colapinto approached him about writing a book about his experience. Reimer agreed to be the subject of a book. In 2000, HarperCollins published Colapinto's book, As Nature Made Him, The Boy Who Was Raised as a Girl. In fact, in 1998, Colapinto published an article of the story for Rolling Stone magazine. In 2000, the article won the ASME Award for reporting.

However, David's life would take a dramatic turn. In 2002, David's brother, Brian Reimer, apparently killed himself, taking an overdose of drugs he was taking for schizophrenia. The death adversely affected David. David spent time at Brian's grave site, mourning his loss.

Despite the royalties of Colapinto's successful book about his life, David reportedly had difficulties making ends meet. Reimer worked at a golf course, doing odd jobs. David reportedly complained that he suffered from depression. On May 2, 2004, Jane told David that she wished to separate. On May 4, 2004, at age 38, David Reimer reportedly committed suicide, shooting himself in the head with a shotgun.

After the news broke that David Reimer had committed suicide, a number of transsexual writers and advocates, including Jamison Green, downplayed the implications of the John/John study. In Green's article on planetout.com, dated May 20, 2004, 'he' compares the David Reimer story to the tribulations of the typical transsexual man's indoctrination that without a penis he cannot satisfactorily function in society as a man. On the contrary the John/Joan study lends weight that chromosomes and the embryogenesis humans undergo more strongly dictate innate sexual identity and not a disquiet about one's gender that in the case of the vast majority of transsexuals materializes in the years of their sexual development. In fact Reimer's story is less of a man troubled by a defect to his genitalia and more of a man troubled by being forced into a social context he could not adapt to. Eventually, Reimer prevails in finding his identity, but by that time it is too late. Reimer is beyond the formative years in learning context sensitive social and gender skills. Reimer also presents as a man somewhat malnourished from the experience and unattractive in virtually every respect.

After learning that David had rejected his female socialization and prior to Diamond breaking the news, when other researchers asked for further information of the famous

John/Joan study, Money was evasive, citing client confidentiality. In fact, Money concealed the fact that Brenda had rejected 'her' gender socialization, after learning the truth at age 14. Despite all his attempts to conceal the truth, Dr. Milton Diamond, of the University of Hawaii, prevailed.

When confronted with the public fall out of the truth of the John/Joan study, Money downplayed its significance in light of his theories of gender, gender identity and the mutability of gender. In fact, Money had a litany of excuses as to why Brenda rejected his gender assignment. Money blamed the timing of gender indoctrination; he said that the indoctrination began too late in the child's development; Money blamed the parents for not following his meticulous instructions; Money blamed the interaction he had with his sibling twin brother. Money, however, could never come to terms with the obvious—the possibility that the theory was in fact fundamentally flawed.

Despite Money's explanations into how John/Joan study went off the rails, Money was reportedly deeply affected. Money was mortified by the case. As a rule Money elected never to speak of it.

* * * * *

As reported in Chapter 2, Sex & Gender, the forces that shape a woman from a man are a series of incredibly complex events that largely occur in the mother's womb. The process of fetal development is the end result of tens of millions of years of natural selection and evolution. Despite all the advances in medicine and genetic manipulation, despite all the advances in plastic surgery, man cannot change a man into a woman nor a woman into a man. Yet, since 1970, eighteen years after the dramatic news of the successful sex change of Christine Jorgensen, the courts began to affect legislation to allow a man to be legally recognized as a woman; and a woman a man. In doing so the courts are not overturning the laws of nature, but rather, they are dispensing compassion towards an otherwise vulnerable and fragile community, respecting self identity above all else. However, recognizing men as women and women as men has human rights implications.

It is no surprise then that when transsexuals assert their legal status on the medical community as something they are intrinsically not and are denied service, these matters test where science and medicine end and human rights begin.

On Wednesday, February 13, 2008 the tribunal of the Ontario Human Rights
Commission (OHRC) heard the opening arguments of a complaint by two male-to-female
transsexuals, Jennifer Finan and Michelle Boyce against plastic surgeon, Dr. Robert Stubbs.
Jennifer Finan sought out Dr. Stubbs for a breast augmentation procedure. Michelle Boyce
sought out Dr. Stubbs for a labiaplasty procedure. Apparently, Dr. Stubbs is an expert at
trimming back labias that become droopy or protruding. The twist is Dr. Stubbs only takes on
female patients and turns away male-to-female transsexuals. Stubbs argument is that he has no
experience treating males regardless of whatever feminizing procedure they may have
undergone. Seems reasonable. Finan and Boyce say not so. [1]

Boyce was the first to speak at the hearing. Boyce, a 38-year-old, coiffed in black curly
hair and projecting an unnaturally high, breaking voice, described himself as intersexed, born
with both ovaries and a penis. Although Boyce grew up male embodied, Boyce reportedly
stated that he always felt female inside. In fact when Boyce purported to be intersexed, he was
being disingenuous with the tribunal. John Money, M.D. of Johns Hopkins University
identified seven classifications of intersexed individuals and none had the combination of a
penis and ovaries. Further, none of the intersexual classifications Money identified are able to
procreate. Boyce fathered two children.

Boyce reported to the tribunal that in 2002 he underwent a GRS procedure in Neeham,
Wisconsin from surgeon, Dr. Eugene Schrang. Schrang gave Boyce a life-like neo-vaginal
cavity, inverting the skin fold of the penis along with the inner lining of the urethra. Boyce
reported that Schrang's work was superficially impressive, however, he also acknowledged
that the surgery didn't go entirely well and one side of the labia minora was bigger than the
other and that he had a protruding flap of skin that otherwise made sex painful.

Boyce, who read about Stubbs' work on women's genitalia, felt he could ameliorate his
post-operative problems and help make sex with his life partner more pleasurable. On
February 20, 2003, Boyce went to see Stubbs for a consultation. On filling out a short
questionnaire on the day of the consultation Boyce wrote down that he was a male-to-female
transsexual and that on October 29, 2002, he had undergone a vaginoplasty procedure. As
Boyce volunteered that he was a post-operative transsexual to Stubbs during the consultation,
Stubbs reportedly invited him to leave. Boyce reported that as Stubbs did so Stubbs was 'rude'
to him. Boyce reportedly chased Stubbs down the hallway and grabbed the doctor and told
him that he could not deny him the procedure. Still, Stubbs refused. Stubbs then tried to

console Boyce, offering Boyce a refund on the consultation. However, Boyce was by this time inconsolable. Feeling rejected, Boyce reportedly cried at the entrance of Stubbs medical clinic—the Cosmetic Surgicentre (Toronto) Incorporated. When Boyce calmed down, he promptly filed a complaint at the human rights commission against Dr. Stubbs that same day. Some time later, Boyce again engaged surgeon Dr. Schrang in Neeham, Wisconsin, but this time he underwent a labiaplasty at a much higher fee than that quoted by Dr. Stubbs.

Boyce's complaint seeks the expense differential between the two surgeons and monetary compensation for 'mental distress'. Boyce reportedly volunteered to Margaret Wendt from the Globe and Mail, who was covering the complaint, that he would be happy to settle the matter for $30,000.00 to $40,000.00.

Finan, who is Boyce's friend, is the other complainant in the matter. Despite being aware of the outcome of Boyce's consultation and despite some trepidation, on March 20, 2003, Finan also went to see Stubbs for a consultation, but for a different procedure. In his case Finan reportedly sought out Stubbs for a breast augmentation procedure. In Finan's case he was reportedly concerned about the size of his breasts and wanted them larger. Unlike Boyce, on filling out a questionnaire Finan did not volunteer the fact that he was a post-operative male-to-female transsexual and not anatomically female. It was only when Dr. Stubbs asked Finan why at age 40 he wanted to undergo a breast augmentation procedure that Finan revealed the truth about himself. On the revelation that Finan was, also, a male-to-female transsexual, Stubbs reportedly became angry and refused to serve him as well. Stubbs reportedly told Finan that he doesn't have the resources or the staff to adequately deal with transsexuals. Stubbs reportedly told Finan that he sympathized with transsexuals and felt that since on October 1, 1998, the Ontario Health Insurance Plan (OHIP) delisted gender reassignment surgical procedures that medical care towards transsexuals in Ontario had become inadequate. In effect Stubbs reportedly tried to console Finan. Stubbs offered Finan a refund on the consultation. However, like Boyce, Finan was inconsolable. Finan reported to the tribunal that what irked him the most was that Stubbs referred to transsexuals as "you people." Finan reported to the tribunal that he was upset and "traumatized" by the experience.

Finan and Boyce first met each other in October, 2000 (or thereabouts) through mutual friends. They each affected a gender transition around the same time and developed a close bond from their respective experiences. In October, 2006, Finan and Boyce went into business together, offering diversity awareness training services. Diversity awareness training services are largely designed to educate corporations of the fact that there are men and women who transgress gender norms and to educate corporations of the implications when an employee affects a gender transition, similar to what Finan and Boyce did. They called their company Diversity Training Live.

Wendt reported Boyce as having feminine gestures. However, Wendt reported that Boyce looks more like a 'guy than a girl'. Wendt reported that Boyce had large hands, course facial hair, a broad-bridged nose and large teeth. Wendt was more impressed with Finan. Wendt reported that Finan passes well as a woman. In her article, dated February 15, 2008, the only knock Wendt had with Finan is that Finan looks like a middle-aged, overweight woman with thinning blond hair.

Stubbs' defence was straight forward. Stubbs testified under oath that he has no surgical experience on transsexuals. Stubbs reported that the chest structure of a male differs from that of a female. Further, Stubbs reported that the post-operative genitalia of a male-to-female transsexual differs from that of a natal female. Stubbs also reported that transsexuals are psychologically fragile and require an entire team of experts and that he doesn't possess the resources to adequately handle transsexuals. Stubbs also reported that he has no prejudice towards transsexuals and similarly refuses to perform chest augmentation procedures on male patients who ask for implants to simulate bodybuilding. Finally, given he is performing elective surgery for a fee, Stubbs is of the belief that he has the latitude to decide who is eligible for his services and who isn't. In fact, Stubbs is duty bound by the College of Physicians and Surgeons of Ontario to turn away people who are in his opinion not eligible.

Over the afternoon Boyce's case reportedly began to collapse. Boyce conceded in cross-examination that Stubbs is in fact not obligated to perform the service Boyce requested, regardless of whether he was qualified or not. After being apparently cleared of wrongdoing from cross-examinations, the only issue left was the matter of Stubbs' alleged rude behavior— Boyce alleges Stubbs intentionally said hurtful things to him and that he (Stubbs) hurt Boyce's feelings. Boyce also tallied in the crying incident outside the offices, which Stubbs allegedly provoked.

Despite the set back from the cross-examination, Boyce reported that the opening day had a profound effect on his life. He reported that he decided to become a "full-time" transsexual activist. Here, too, Boyce is being disingenuous. In fact Boyce had started to advertise himself as an advocate and a champion of human rights as early as 1999.

The timing of Finan's visit to Dr. Stubbs' medical office to inquire about a breast augmentation procedure is also suspect. Finan reported that although he was genuinely interested in undergoing a breast augmentation procedure and he had an interest in Dr. Stubbs due to his reputation in the field, he did admit that Boyce manipulated the situation to serve his own interests when the matter eventually came before the tribunal.

In any event the matter did not conclude on February 13, 2008 and in fact the matter stretched out to six separate court appearances over a five month period. The strain of the court appearances had reportedly taken a toll on Finan and Boyce. In March, 2007, Finan sold off his equity interest in Diversity Training Live to Boyce. Simultaneously, Finan also terminated his personal relationship with Boyce, refusing to speak to him. Finan refused to comment on the reasons. Although Finan and Boyce attended each hearing, they had no appreciable interaction.

Finally, on July 22, 2008, adjudicator David Wright, who heard the matter on behalf of the Human Rights Commission of Ontario, dismissed the complainants' claim against Dr. Stubbs. [2] In his endorsement Wright recognized that Finan and Boyce are legally and socially women. However, Wright was more persuaded by the evidence put forward by Stubbs that the complainants are physically different from other women from a medical perspective. In the absence of contradictory expert evidence Wright found Stubbs' handling of Boyce and Finan as responsible. He writes:

> For professionals, knowledge of the limits of one's own expertise and skills is an important part of good practice, fulfilling professional obligations, and serving the public adequately.

Further, although Wright appreciated the complainants' argument that transsexuals, like them, face significant barriers to access to health care and that the health care system has in effect failed them, Wright had grave difficulty in finding fault in one cosmetic surgeon, who chose not to seek further qualifications to accommodate that gap and whether that marks a violation of the Code. He writes:

> I find the accommodation that the Commission and the complainants seek, namely that Dr. Stubbs obtain significant new qualifications and training and change the nature of his practice, would have amounted to undue hardship.

Finan who was disappointed with Wright's ruling reported that in fact, the OHRC didn't do due diligence in bringing forward expert opinion to demonstrate that Stubbs' lack of knowledge of transsexual care was reasonable. In fact Wright admitted in his endorsement that he relied completely on the testimony of the three parties and did not hear any expert opinion.

"I talked to a number of plastic surgeons about [a] breast augmentation [surgical procedure] and they all say that the [procedure] is the same whether it's done on a natal woman or transsexual woman. Stubbs never made the effort to investigate whether performing the surgery on a transsexual woman matters." Finan says.

Finan reported that the complaint was an uphill battle and that the OHRC did little to serve the interests of transsexuals. Further, Finan reported that the OHRC intentionally didn't seek expert opinion and even if they had, they would have run into political obstacles.

"Doctors won't testify against other doctors. It's that simple." Finan says.

Human Rights complaints move at a snail's pace. The complaint process begins with the complaint, itself. The respondent must be notified and he/she must prepare a defence. The two parties then engage in mediation. If mediation doesn't work, the tribunal reviews the case for merit. If the tribunal believes the case involves a serious question, the tribunal hears the complaint. Although the Ontario Human Rights Commission appoints a lawyer to represent the complainant, the respondent is left to his own. In this case, Stubb's medical insurance is covering his legal expenses. However, given medical practitioners in Ontario are under the Ontario Health Insurance Plan (OHIP), from any angle you look at Finan and Boyce versus the Cosmetic Surgicentre (Toronto) Incorporated and Dr. Robert Stubbs, the Ontario tax payer is picking up the tab in its entirety.

*　*　*　*　*

The current psychiatric opinion on transsexualism is that the men and women, who exhibit a so-called 'disquiet' about their birth gender, have what is known as gender identity disorder. Gender identity disorder is the discomfort one has with their assigned sex at birth coupled with the desire to be a member of the opposite sex. Psychiatrist Paul McHugh of Johns Hopkins found that the disorder naturally occurs in males, who are assigned as members of the opposite sex, at birth. McHugh cited David Reimer as an example of a male reassigned at birth, having a genuine gender identity disorder.

The modern research suggests that unless the child is disquiet about their assigned gender prior to the age when the child begins to sexually develop, gender identity cannot be the cause of gender variant behavior. Males begin to sexually develop at age five. Unfortunately, the American Psychiatric Association that governs the Diagnostic Statistics Manual, has no other explanation for persistent gender variant behavior, leaving a gaping hole in the science behind gender variant behavior and the life stories of transsexuals. No where in the write up of the diagnosis is there any mention of a sexual response underlying the motivations for seeking a gender reassignment surgical procedure.

Prior to 1991, the landscape is in absolute darkness in the understanding as to why some people appear on the surface to benefit after undergoing a GRS procedure (or stay the same) while others fail miserably to benefit from undergoing one.

* * * * *

Lynn Conway, a professor emeritus of electrical engineering and computer science at the University of Michigan at Ann Arbor, publishes a page that includes a number of individual profiles illustrating "successful" post-operative male-to-female transsexuals. Lynn Conway, who is not a researcher, provides the topic of transsexuals success stories as a community service and hopes to inspire and encourage other people, who exhibit transsexual proclivities, to undergo a gender transition, involving a GRS procedure. Conway is 'herself', a male-to-female transsexual. In fact, Lynn Conway, who looks remarkable given her age, appears in a number of the images of the subjects of the so-called success stories. Conway is now seventy.

Conway does not publish the criteria of success other than to state that the subjects have all 'successfully' assimilated in society as women. Even with this narrow criteria a number of the profiles that Conway lauds as 'success stories', are questionable. The only apparent thing the individuals in the profiles have in common is that they all reportedly underwent a gender reassignment surgical procedure and now live as male-to-female transsexuals in society, like Conway. In fact, the case studies are poorly researched and include a number of misreported facts. Further, a number of profiles include links where the pages no longer exist on the World Wide Web.

I define success insofar as a gender transition to include a number of criteria. The criteria below are somewhat more rigorous than Conway's criteria.

1.	career retention: the individual continues to engage in the same career after transition as they had enjoyed prior to a gender transition. If the individual loses their job due to a gender transition, the individual is able to re-emerge in the work force at the same level of responsibility;

2.	family retention: the individual continues to have a positive relationship with close family members, including their parents, siblings and any children; and

3.	relationship retention: the individual continues to engage in friendships outside of the community of transsexuals.

When you apply the criteria above very few of the subjects appearing on Conway's page are successful, if any.

Mianne Bagger

Conway reports Mianne Bagger as a TS success story. Conway credits Bagger as a "professional golfer." The innuendo is that Bagger is a successful professional golfer and transsexual. [4]

In fact Bagger is not a successful professional golfer in the context of her anatomical sex. Bagger currently plays on the Europe Ladies Tour. Prior to his gender transition, Bagger never qualified as a professional golfer as a male. Bagger only realized a career in professional golf after undergoing a gender transition and only in competition with females.

While it is true that Bagger is a card carrying member on the Ladies European tour, and while it is true that in 1999, Bagger won a female amateur championship, Bagger is not a successful tour player on the LET. Over the course of the 2004 season, Bagger, who was 38-years-old at the time, failed to make a single cut on any tournament event. The LET reports that Bagger's career earnings for 2004 as $0.00. Over the course of the 2005 season, Bagger played in 13 tournaments and earned Euro$2,726.00. Over the course of the 2006 season Bagger played in 8 tournaments and earned 17,385.83. Over the course of the 2006 season, Bagger's average earnings were Euro$2,173.23. Over the course of the 2007 season Bagger played in 16 tournaments and earned 39.265.41. Over the course of the 2007, season Bagger's average earnings were Euro$2,454.09. Bagger is currently ranked 52 on the LET.

Mianne Bagger earned a total of $59,278.32 of gross income over four years on tour. Bagger earned an average of Euro $1,347.00 of gross income per tournament. If you account for the expense of flight and accommodation, you quickly discover that Mianne Bagger is in all probability earning negative income after each successive tournament.

Bagger reports the Hillerod Golf Club as a sponsor. Bagger also reports the site, golfonline.dk, as a sponsor. Bagger also reports her family as a sponsor. Bagger does not publish any information that would suggest she is earning any money through endorsements. Generally, tour players on the LET, who rarely crack the top ten, are not the media darlings of sponsors. Although Bagger's earnings are progressively higher year after year, she is no longer at an age where one would expect a birth in the LPGA. Bagger is now 41-years-old. Bagger's chances of making it onto the LPGA (where the money is) are remote.

Despite Bagger's argument that she belongs in the Ladies competition due to the fact that she is legally female, Bagger ignores the fact that she benefitted from over twenty formative years of the testosterone levels of a normal male. If she was female, the testosterone levels of a normal male would otherwise be a performance enhancing drug and would discount her from any competition where the sport regulated drug use. Bagger rebuts this fact by arguing she is ranked as 163 in driving distance in 2006.

In effect Bagger is living a 'dream', but she has not earned the 'dream'. Bagger is in all likelihood not earning a significant income pursuing a career in professional golf. Since her entry into the LET in 2004, Bagger has received little media attention for her accomplishments in golf.

Bagger reports that during his gender transition he was consumed by gender dysphoria and didn't attempt to compete with men for that reason and for that reason only.

Christine Beatty

Conway reports Christine Beatty as a TS success story. Conway credits Beatty as being a software engineer and a writer.

Beatty reported that he had a long history of drug addiction and that he eventually became a heroine addict. Beatty also reported that while living as a man, he had a PCP induced psychotic episode. Further, Beatty reported that he found himself in jail at one point. Further, Beatty reported a number of failed attempts at rehab. Further, Beatty reported fresh "disasters" in his life after undergoing an orchiectomy procedure. An orchiectomy procedure otherwise protracts the testes.

Beatty reported that he formed and played in a transsexual music group, but the band broke up. Beatty cited the breakup was due to the dysfunctional lifestyles of the band members, who were also transsexual. However, Beatty reportedly continued to play music and formed what were known as Glamazon. Beatty reported that he played with the group, while retaining a job in software. In 2001, Beatty reported that he quit Glamazon.

Beatty reported that he was invited to speak on a number of radio stations and panels about his transsexuality. Beatty reported that he received an award from the community of transsexuals for his activism.

It is unclear what role, if any, a gender transition or a GRS procedure had in his ability to overcome a drug addiction.

Beatty has published a number of videos of himself on YouTube. When you watch the videos of Beatty it is hard to imagine that Beatty passes as anything other than a male-to-female transsexual, who is a recovering drug addict.

Melanie Anne Philips

Conway reports male-to-female transsexual, Melanie Ann Phillips as a TS success story. Conway credits Phillips as being a "TS Support-Site Authoress".

Phillips, who reportedly first began publishing on the World Wide Web in 1994, is one of the early adopters of the new self-publishing media, based on the HTTP protocol. Phillips is the mind behind the transgender support site on the domain, heartcorps.com. Since embarking on his gender transition in 1987, Phillips has published volumes of material about his transsexual experiences, shifting from Dave to Melanie. In fact, to date Phillips has authored a diary of 106 chapters, covering five volumes of material, beginning with the first installment, Point of Departure, dated October 3, 1989. The material includes copy and audio material. The transsexual diary largely covers his eighteen-year transsexual experience interalia. An interest in transsexualism (or becoming female) is not indigenous to females. 99.99% of adult females have no interest in the topic in men or women whatsoever, but Phillips—who subscribes to the female narrative of the female-trapped-in-a-male's body syndrome —reports he is otherwise a typical woman were it not for his being born male.

In addition to the transsexual diary Phillips also publishes a webzine, <u>The Subversive,</u> which he no longer contributes to. Phillips produced a total of 36 editions of <u>The Subversive</u>.

In addition to his transsexual diary and the webzine Phillips also publishes a large photo album of himself in all various stages of transition, including images of himself prior to his gender transition, when he lived as Dave. In one image Dave, a.k.a., Melanie is sitting in a tub mustachioed. Phillips, a.k.a., Dave, who was married at one time and has two children, is otherwise heterosexual. Phillip's children are now of adult age.

In his photos taken early in his gender transition and prior to undergoing the GRS procedure Phillips looks 'cute' save the image where he is wearing a bikini, which he has trouble pulling off. The images are scanned from film photography and are not of the same quality of digital photography. If digital photography were available at that time, Phillips may have had more trouble pulling off the 'girl next door' look. Further, in a number of the images from Melanie's dairy he is wearing sexually provocative lingerie, including a pink teddy. In yet another he is seen leering at his breasts—exposed to the camera—and squeezing them. In the caption he writes:

> I think there's something very sensual about a woman touching herself. From the model's point of view, at first I felt a little uncomfortable with this. Posing nude is one thing, but touching your own naked body is quite another. But, after seeing a few of the shots, I realized how much I looked like the girls in the centerfolds that I had always wanted to grow up to be.

In any event Phillips, who reportedly underwent a breast augmentation procedure, is being disingenuous with his audience. They aren't real nor spectacular.

In addition to the transsexual diary, the webzine and the photo album, Phillips publishes a self-help video and audio set, Develop a Female Voice. The product set is available on Yahoo Stores—an online order entry system. On the site Melanie advertises that he has otherwise mastered the art of the female voice. When Melanie is 'on' he does have natural sounding female voice—it even comes over sweet—however, on a number of videos where he appears off guard he comes across with a voice typical of that of a transsexual. Melanie reports that he has not undergone any surgical procedure, affecting his voice, which is a selling point for the material.

Phillips was reportedly born in 1953, and began his gender transition in 1989 at age 36. Phillips reported that he underwent a GRS surgical procedure in 1992—three years after undergoing hormone replacement therapy (HRT). Again, Phillips reported that at one time he was married and fathered two children. However, Phillips marriage ended at the time he underwent the GRS procedure.

Phillips' proclivities towards feminization apparently did not stop with the GRS procedure in 1992. Phillips reported undergoing a facial feminizing surgical procedure in 2006. Phillips reported that the surgery was painful, but that he was pleased with the result. Phillips says the reason he underwent the procedure was to deal with a problem with his skin, which succumbed to aging. Phillips was 53-years-old at the time he underwent the procedure.

In real life Phillips sells software online to help writers create story lines and write screenplays. Phillips also sells Online Kits to kick start people, who are aspiring writers.

references

[1] GLOBE & MAIL ARTICLE, A DAY AT THE THEATRE OF THE ABSURD

On February 15, 2008 the Globe and Mail newspaper published Margaret Wente's article, A Day at the Theatre of the Absurd, featuring the human rights complaint, Jennifer Finan and Michelle Boyce versus Cometic Surgicentre (Toronto) Incorporated and Dr. Robert Stubbs.
more information
visit the site: Globe and Mail newspaper article, A Day at the Theatre of the Absurd
http://ago.mobile.globeandmail.com/generated/archive/RTGAM/html/20080215/wcowent16.html

[2] THE HUMAN RIGHTS TRIBUNAL OF ONTARIO DECISION IN THE MATTER OF JENNIFER FINAN AND MICHELLE BOYCE V. COSMETIC SURGICENTRE (TORONTO) INCORPORATED AND DR. ROBERT STUBBS

The Human Rights Tribunal of Ontario Decision on the matter of Jennifer Finan and Michelle Boyce versus the Cosmetic Surgicentre (Toronto) Incorporated and Dr. Robert Stubbs is available online.
more information
Jennifer Finan and Michelle Boyce vs. the Cosmetic Surgicentre (Toronto) Incorporated and Doctor Robert Stubbs (Tribunal Decision)
http://www.cup.ca/jhm/hrto.pdf

[4] LYNN CONWAY PROFILE, MIANNE BAGGER

Lynn Conway features Mianne Bagger as a transsexual success story and a successful professional.golfer on the Ladies European Tour. Although the LET does list Bagger as a card carrying member, the site has very limited information on Bagger.
more information
visit the site: Baggar's profile on Conway's site
http://ai.eecs.umich.edu/people/conway/TSsuccesses/TSgallery3.html
visit the site: Ladies European Tour web site

http://www.ladieseuropeantour.com/
visit the site: article on Bagger
http://www.smh.com.au/articles/2004/02/14/1076548274856.html

CHAPTER 04

Men Trapped in Men's Bodies

In 1991—eleven years after the American Psychiatric Association introduced Transsexualism as a distinct psychiatric disorder [DSM-III, 1980] only to rearticulate the disorder as Gender Identity Disorder (GID) seven years later [DSM-III-R, 1987]—a champion in the understanding of cross-gender behavior emerged. In 1991 psychologist Ray Blanchard of the Centre for Addiction and Mental Health (CAMH) published his first article into autogynephilia. Autogynephilia—as Blanchard coined it—is the propensity of a male to become aroused by the thought of himself being or becoming a woman. The word, itself, derives from the Greek language, meaning "love of oneself as a woman".

Blanchard's work into autogynephilia is nothing short of a breakthrough in the understanding of transsexual behavior. Blanchard's insights into transsexual behavior are unquestionably brilliant and a complete departure of the accepted understanding of transsexualism at the time.

Blanchard, himself, is the most unlikely of heroes. Blanchard is reportedly diminutive, cynical, and insensitive. Blanchard is also reportedly not compassionate. However, in what he lacks with style and charisma he more than compensates with analytical and deductive reasoning.

In 1991, Blanchard, who went on to become the head of the gender identity clinic of the Centre for Addiction and Mental Health (CAMH) (formerly the Clarke Institute of Psychiatry), identified a topology of transsexuals, delineated by sexual orientation, including homosexual, heterosexual, bisexual and asexual. Blanchard's concept of autogynephilia effectively lumps the three subtypes, heterosexual, bisexual and asexual males, as autogynephilic.

Autogynephilia can exist in otherwise homosexual men. In this case the male's masturbation thoughts involve functioning as a female during intimacy with a male and not necessarily the male object preference. Blanchard classifies this subtype as asexual or having no sexual interest in third parties. If you include otherwise asexual males there are virtually no natural cases of homosexual, male-to-female transsexuals, who do not have autogynephilia to one degree or another.

Blanchard's work does not necessarily suggest that autogynephilic transsexuals always masturbate to the thought of being a member of the opposite sex. Rather, Blanchard suggests that the masturbation thoughts that began at a young age in a male child's sexual development steer the person towards a gender transition later in life. The person effectively suffers from compulsive thoughts to undergo procedure after procedure in an effort to realize the established masturbation thought. The compulsion is not unlike that of transvestic fetishism in men or periodically masturbating, while wearing women's clothing.

The self realization of autogynephilia effectively runs counter to the person's self constructed mental fantasy and delusion of being a female trapped in a male's body. The autogynephilic male rejects any external suggestion that strays from the accepted transsexual narrative—that he is otherwise a woman born with a birth defect of having male genitalia. Further, the autogynephilic male punishes any person, who articulates the suggestion as it violates his self constructed fantasy and delusion about himself. Blanchard's work states that the nature of the disorder, itself, which is sexual and irrepressible in nature, compels the person affected to reject contrasting opinions and anyone associated with contrasting opinions. A psychiatrist can no more convince an autogynephilic transsexual that he is autogynephilic than a psychiatrist can convince an anorexic that he or she has an eating disorder and is dangerously under weight.

Unfortunately, at the time Blanchard put forth his findings into cross-gender phenomena and autogynephilia, the political landscape surrounding transsexualism and the treatment thereof was well developed and entrenched. In 1991, transsexual patients were already well organized. The community of transsexual patients overwhelmingly rejected the suggestion or finding that a sexual response is the underlying motivation for cross-gender proclivities. The medical industry serving transsexual patients was also well established. Plastic surgeons, who honed their skills towards altering people's sexual anatomy, and operated private clinics, were not interested in abandoning their multi-million dollar practices to effectively start over.

Psychiatrists, who evaluated and approved men and women for gender reassignment, were also not interested in abandoning their positions as industry experts. When Blanchard, Steiner, Clemensen and Dickey published a study into postoperative regrets that demonstrated a strong correlation between sexual orientation and the propensity of explicit postoperative regret, their warnings of problems in heterosexual male transsexuals largely fell on deaf ears. The study was quantitative in nature and not qualitative. The authors did not personally interview the sample of postoperative transsexuals to know how the individuals are in fact performing in society as members of the opposite sex. [1] Leonard Clemensen has since undergone a gender transition and has since emerged as Maxine Peterson.

The concept of autogynephilia had not received much attention outside of the field of sexology until Dr. Anne Lawrence—a male-to-female transsexual, medical doctor and self-identifying autogynephile—published a series of articles about the concept in the late 1990's. The articles are available online at her web site. In fact, in 1998 Anne Lawrence was the first to coin the term, 'men trapped in men's bodies' to describe the vast numbers of male-to-female transsexuals, who did not fit the stereotypical transsexual for which gender identity disorder was articulated for. The autogynephilic transsexuals that Lawrence described as men trapped in men's bodies, who desperately wanted to be women, all had a common background that was different from the stereotypical background of an apparent homosexual transsexual. In the case of the heterosexual transsexual, Lawrence reports that these transsexuals often experience successful careers as men, were at one time married, and demonstrated a low degree of femininity.

* * * * *

It is 9:00 p.m. at David's house in the Danforth Village of Toronto, Ontario on Saturday, January 29, 2000. David, who recently separated from his wife and now lives in the matrimonial home by himself, has a number of guests over. All of the guests are male-to-female transsexuals. David (not his real name) and the group are getting ready to go out. In the case of the male-to-female transsexuals, who are otherwise straight men; many of whom are involved with women, getting ready involves hours of shaving, plucking, painting cosmetics onto one's face, dressing and accessorizing. Depending on how far the transsexual is in his gender transition, the process can take anywhere from twenty minutes to two hours. Tonight, David and the 'girls' plan to go to Pimbletts, an old English tavern, for a beer or perhaps two. Pimbletts is notorious to cross-dressers and transsexuals across North America. After that David and the 'girls' plan to go to Zippers—a gay dance club—to end the evening on the dance floor. Daniel, a.k.a., Danielle is among the group.

Like many men and women, Daniel's odyssey into transsexualism reportedly began in September, 1996—at age 26—as he explored the Internet for pages, having relevance to transsexualism. However, unlike many men and women, Daniel (not his real name) thought what he had stumbled onto as a child was unique and special and that it belonged to him and him alone.

"I regarded my own cross-dressing as something that made me original. It was my idea and my idea alone. Until that time I didn't know anyone that wanted to dress in women's clothing. I didn't know anyone that wanted to be a girl. I thought it was my little secret and my little life. I felt special." Daniel says.

"Once I discovered all those web sites and the existence of the transgender [community], I went into a state of depression. I was so upset knowing other people did this I refrained from reading their sites."

In September, 1996, at the time Daniel began to explore the Internet for materials on transsexualism, he was the married father of one child age 2.

Daniel met his first wife, M., in 1989. Daniel was 19 years old at the time. He proposed to her six months later. In 1992, three years later, the couple wed. Daniel and M. had their first child, D., in 1994 and their second child, J., in 1998. Daniel reported that M. was otherwise a virgin when he met her. The family lived in Aylmer, Ontario, which is near London, Ontario.

During the marriage, M. worked as a registered nurse. She typically worked 12 hour shifts twice per week. She financially supported the family by paying the mortgage, bills and household expenses. Daniel had an income from a computer business however, all the profits were reportedly re-invested into the business.

Daniel reported that his wife had a sexual fascination in male/male sexual acts. Female interest in male/male sex acts is not unusual and is analogous to men having an interest in lesbian sex acts. The interest is otherwise innocuous.

Daniel's bisexual curiosity and M.'s interest in male/male sex acts intersected in their marriage. The couple practiced an 'open' marriage.

"The ground rules were if I met a guy I liked, that was fine. However, I was never allowed to touch another woman. In fact, I went on to have sex with a few other males while I was married. I would meet guys and bring them back to our home to have sex. My wife would casually watch me having sex with other males. It was fun. [Her participation in watching me have sex with other males] didn't do anything for me per se." Daniel says.

"She and I didn't have much of a sex life together. This gave me a sex life and from watching me she got turned on. That enabled us to have a sex life." Daniel says.

However, M.'s interests in male/male sex acts did not extend to cross-dressing. Daniel kept his proclivities towards cross-dressing private from his wife. When he cross-dressed, he did so in private.

"My only opportunity to cross-dress would be when my wife left for work. She typically worked at night. So in the evening after she left for work, I changed into women's clothing. I would set the alarm at six o'clock in the morning to change back before she returned home so she wouldn't see me. At times I wore a mustache to keep myself under control."

In 1997, Daniel's interests in transsexualism began to escalate. One Saturday in September, 1997, Daniel, who was still camouflaged behind a mustache, went to White Oaks mall in London, Ontario. He walked into a wig shop and purchased a wig on the auspices that he needed the wig for a cross-dressing party. Daniel reported that his hair was otherwise short at the time. Daniel selected a wig with long wavy black hair, which is similar to his own when grown out. When Daniel returned from the mall he shaved off the mustache. After the mustache became a fading memory, Daniel proceeded to take film pictures of himself in the wig.

"I blew off a couple rolls of film." Daniel says laughing. "I ran out and had them developed to work on my look."

The next Friday, after the children were in bed, Daniel dressed himself in one of his wife's tops, his wife's blue jean skirt and blue high heels. He then engaged his wife and told her that he was leaving the house to pick up milk at a Becker's convenience store in a nearby town (population 4,000) dressed the way he was. M. objected, pleading with him to reconsider, but Daniel left, wearing his wife's clothing, anyways. When he arrived at the Becker's convenience, his nerves got the better of him. He stayed in the car, until he calmed down. Daniel, then, with great determination proceeded to enter the store, grabbed one of the cartons of milk from a large fridge and went to the cashier to pay. What happened next was perhaps the most underwhelming moment in Daniel's life to that point. The clerk reportedly engaged Daniel in conversation without any hint that anything was out of the ordinary—or so he thought.

The second time Daniel went out as Danielle, he went on an impromptu visit to see a boyfriend—Michael. In fact on the visit he took his daughter, J., with him. J. was less than a year old at the time. Michael lived on St. Clair Avenue in Toronto, Ontario. Daniel reported that when he arrived at Michael's house, Michael wasn't home, so to pass the time he parked the car in front of an adjacent park and waited for Michael to return. Time passed. Eventually, Daniel grew tired of waiting and drove his daughter and himself to a mall in Kitchener, Ontario, where he reportedly went on an eight-hour shopping spree. Kitchener, Ontario and Toronto, Ontario are ninety minutes apart by car, or thereabouts.

"It was a riot." Daniel says.

Glowing from the shopping spree, Daniel immediately went back to Michael's place on St. Clair Avenue to show Michael how passable he was as a woman. This time Michael was home. However, Michael didn't see Danielle as intriguing or as passable as Daniel saw himself.

"He didn't like me that way. It completely changed our relationship. He couldn't see me as a guy anymore. I saw him over the course of a couple more dates which dragged things out for months, however, that experience ended the relationship." Daniel says.

That same year Daniel reportedly began to publish materials, advertising himself as a transsexual sex worker. He reportedly earned small amounts of money through sex work. As he did so he discovered his first wife, M., was involved in an extramarital affair.

Given the tangential nature of the sexual relationship between Daniel and M., given the 'open' marriage pact Daniel and M. had effected, Daniel didn't object to his wife's extramarital affair. In fact he encouraged it. M. romantically began to see the man, who lived in Atlanta, Georgia, that she interactively chatted with over the Internet. Due to the great distance between the two regions, the relationship was doomed to failure. M. reportedly only made the trip to Atlanta, Georgia one time.

Between March, 1999 and November, 1999 Daniel began to collect a sizable wardrobe of women's clothing and paraphernalia for the purposes of gender reassignment. M. reportedly resented the expense, but continued to support the entire household. Eventually, in November, 1999, M. moved to St. Thomas, Ontario with the help of her family, abandoning the matrimonial home in Aylmer, Ontario.

Daniel only finally connected with another transgender person like himself on a trip to Las Vegas. There he met Barb—an intersexed person reportedly —whose parents had assigned as male as a child, but was undergoing a gender transition at the time. Barb worked as an electronics specialist at a casino. Daniel reported that Barb was bright and capable. However, when Barb affected a gender transition, his life reportedly got worse and not better.

"[Barb's] decision to transition as a transgendered individual; her goals; pretty much cost her everything in life, including her marriage, career, house." Daniel says.

Barb ultimately declared bankruptcy and in the process put himself in a position where he could no longer qualify for a bond. Without the ability to take out a bond, Barb could no longer work at a casino. He effectively became homeless. Prior to Barb's collapse, he operated a web site out of Las Vegas. The site was reportedly very popular and included self-help material and other resources geared towards gender transition, but when Barb moved into the back seat of his car, the web site closed.

Despite seeing the living, breathing consequences of what can happen to a technically gifted individual, who affects an ill conceived gender transition, Daniel failed to benefit from the experience. Like Barb, Daniel was at the time an aspiring self employed, information technology professional, trying to break out. He was a natural-born leader. Daniel reported that he otherwise worked eighteen-hour days and did little outside work and sleep. Perhaps eighteen-hour is an exaggeration.

In any event, witnessing Barb's collapse reportedly had a dramatic effect on Daniel. Daniel came to the belief that people matter, but more than that, he also came to identify as a transsexual advocate and a champion of human rights and particularly the rights of transsexuals—what he saw as the most vulnerable of communities. Daniel's advocacy and self identity of being a champion in effect both bound him and engulfed him in transsexualism, which underneath the rhetoric served his own political agenda and the plans he set down for his own body. That one singular purpose ultimately propelled him to closely follow in Barb's footsteps and marked the critical flaw in reasoning he would never recover from—personally or professionally.

"What that triggered in me is that I felt something politically had to be done and that this couldn't be allowed to happen. Shortly after that I began a crusade. The crusade is transgendered rights; to be accepted in society as a transgendered person and to not lose your rights and status in society by making that statement." Daniel says.

"The term, transgenderism, itself, is an umbrella terms that describes a lot of people. However different these people are, they all have the same major issues to deal with, that being acceptance in society, access to washroom facilities, and the right to be and function as

58

themselves without discrimination. That's my goal in life and I have worked very, very hard at it."

He did, tirelessly so. In any event the world didn't need another transsexual advocate or champion of human rights. There are plenty of those to go around.

Transsexual advocates are a funny breed. Male-to-female transsexual advocates tend to erect large and expansive web sites to educate people on all facets of transsexualism, including what hormone replacement therapy is, human rights with respect to transsexuals, work place issues, essays on the need for the state to subsidize all surgical procedures from GRS to facial feminization procedures, all complete with before and after photographic images. These sites tend to advance the owner's emotional needs and not necessarily the needs of the community at large in any cohesive manner. Once the owner undergoes an invasive feminizing procedure, like a gender reassignment surgical (GRS) procedure or similar procedure that protracts the testes, the owner's interests in advocacy tends to evaporate and the site either sits in abeyance or the owner closes it to save money.

Rosalyn Forrester, a.k.a., Harold Forrester, closed his self-advertised advocacy site, Canadian Transsexuals Fight for Rights, at the domain, ctffr.org, around the time he underwent a GRS procedure. Daniel's own site, that advertises diversity awareness training services, is antiquated in terms of graphic design. The site is authored in Microsoft FrontPage 5.0—an HTML authoring tool—which Microsoft has since discontinued. None of the dates published on the site is more current than January, 2006. In fact the only document on Daniel's site that appears current is the PDF brochure, which lists ten client engagements—only three of which are corporate seminar events.

As his attention began to narrow onto transsexualism, he became financially irresponsible. Daniel reportedly collected provincial sales taxes for his business but failed to remit it to the government. The Ontario government reportedly suspended his vendor license due to unpaid fines.

Daniel became involved in the transsexual community in Toronto, Ontario in July, 1999, when he became involved with David, his eventual life partner. Daniel met David on an Internet chat site. The two have had an on-again off-again relationship ever since. Daniel also participated on the collaboration services of the site, Take a Walk on the Wildside. Although Daniel had no interest in the fetish boutique, Take a Walk on the Wildside, nor the proprietor, he didn't take exception that most of the people who interactively communicated on the site, wildside.org, were cross-dressers. He tolerated their sexual innuendos. He also organized social gatherings where he got to meet more and more people like himself.

In fact David made Daniel's transition feasible.

"[David] has given me an environment to transition freely and without obstacles. When I come to Toronto and I'm here with my friends, basically, [David] is affording me a second life. It has provided me with stability. Probably too much stability. I tend to take advantage of him a little bit too much. But he has provided me with a place to be myself for long periods of time, three days here, five days there. I don't have to worry about a school meeting where they don't know about me, for example. Then I slip back into practicing all of the hiding traits."

Eventually, Daniel told his mother that he wished to affect a gender transition. On hearing the news, Daniel's mother blamed herself for Daniel's proclivities towards transsexualism. She reportedly had heard rumors that going off the pill just prior to pregnancy caused abnormal sexual deviancy in male offspring and felt her action of going off the pill just prior to getting pregnant with Daniel affected his sexual identity. Daniel assured his mother that he liked being transsexual and that transsexualism gave his life meaning and if she did something to cause that then he felt grateful.

By April, 2000 Daniel's gender transition was moving at breakneck speed. Although he lived in Ontario, Daniel applied to have his name changed in Alberta. Daniel discovered that any piece of identification that bears the gender code, 'F', helps to affect changes to other pieces of identification, including one's birth certificate, which is what Daniel sought.

In 2000, Daniel was constantly travelling back and forth from Toronto, Ontario to Aylmer, Ontario to meet his parental obligations. He brought his children on a number of these visits.

In only two years from emerging in the transsexual community of Toronto, Ontario as a visible transsexual, in October, 2002, Daniel underwent a GRS procedure with Dr. Eugene Schrang in Neeham, Wisconsin. However, the surgery and the subsequent recovery period, which ran through December, 2002, reportedly did not go well and Daniel reported that urination and sexual intercourse were painful. In addition to the pain, Daniel also reported he did not care for the physical appearance of his neo vagina. Daniel also reportedly had concerns about the post surgical care with Schrang. In any event Daniel reportedly returned to Neeham, Wisconsin again and again, spending more and more money, which his wife reportedly gave him, on his gender transition.

On Daniel's return the household in Aylmer, Ontario began to dramatically deteriorate. M. often attended the residence in Aylmer, Ontario to clean the house and interact with the children. She reportedly put them to bed. However, when M. was no longer willing to clean the house, it became filthy. On the event a social worker from the Office of the Children's Lawyer visited the property, Daniel cleaned the house. He took pictures of the property for an attendance at court. However, Daniel reportedly relied on his partner, David, to clean the house.

Daniel also reportedly ran into legal troubles. Daniel and his estranged wife, M., engaged each other in a long protracted family matter over custody of the children, stretching over two-years. Daniel reportedly sought joint custody with respect to the children, which his wife contested. The family matter became ugly and in the end Marshman J., who heard the matter, sided in M.'s favour, stripping Daniel of his custody rights to the children.

Marshman J. had a number of concerns regarding Daniel in the light of a custodial parent, including Daniel's insistence that the children know intimate details of his gender transformation, including the various surgical procedures he had undergone and dilation—the act of applying stents of different sizes to the surgically created vagina to keep it from closing. In fact if a post-operative male-to-female transsexual fails to apply the stents to the reconstructed genital region, the surgically constructed vagina will close like any other non-indigenous wound to the male body. Marshman J. came to the opinion that Daniel's interests in transsexualism were not necessarily in alignment with the best interests of the children. Marshman J. also had concerns about Daniel's apparent lack of responsibility. He found that Daniel was far more interested in the status of motherhood than in its responsibilities. In his endorsement, Marshman repeatedly cites that Daniel provided little stability and frequently moved the children between residences. He found this practice adversely affected the children's attention to their respective homework. In his endorsement, Marshman J. also cites that Daniel frequently returned the children to their mother late. Marshman J. also found Daniel to be evasive throughout the legal proceeding. In his endorsement, Marshman J. cites as example that Daniel testified that his sister had always been involved in his children's life, then minutes later admitted that he had not spoken to his sister for four years. Marshman J. also had concerns about a 56-year-old, male-to-female transsexual drifter, named Dianne LaLonde, who Daniel invited to live with him at the residence in Aylmer, Ontario for eight months and the effect the arrangement had on the children. In his endorsement, Marshman cites that D., Daniel's son, didn't like LaLonde, but that his objections didn't seem to faze Daniel. LaLonde was at the time on a suicide watch and had attempted suicide on at least one occasion while living at the residence. Here too, Daniel's acts of charity towards a fellow transsexual and his insistence that people matter—apparently more so than his own children—only served to cost him and in the end it cost him the joint custody of his children if not in whole, then in part.

Daniel reported that his transvestia began in early childhood; around age 4; around the time young males sexually develop.

"When I was four my best friend was a girl my age named, Michelle. She lived down the street. She was a Tom boy. Our respective mothers took turns baby sitting us. We spent all of our time together just the two of us. One day we realized we had something in common. Michelle told me that she didn't like to wear clothing for girls and wanted to wear my clothing. I told her that I didn't like to wear clothing for boys. So we exchanged our clothes." Daniel says.

"The first time it happened I spent the night over at her house. I didn't have a pair of pajamas. So I wore one of her nightgowns. Later, as I would spend the evening sleeping over at her house I would routinely forget to bring my pajamas. I felt more comfortable as a girl than as a boy. At that time I had no idea that there was anything wrong [with] wearing girls clothing."

Unfortunately, disaster struck. Michelle's mother caught Daniel in a pair of her daughter's panties under the porch of her house, but the woman never relayed the news of the 'abhorrent' act to Daniel's mother.

"After that incident Michelle and I continued to exchange clothing, but we were much more careful. Then a few years later my sister [who was two years younger than me] came along and I began to wear her clothes. When I turned nine I began to take pictures of myself dressed as a girl. At this time I began to sneak out of the house dressed as a girl. I took long walks through the neighborhood when I was supposed to be in bed." Daniel says.

Daniel reported his bisexual tendencies began in early adolescence. He reported having a sexual encounter with another boy his age.

"I was playing with a male friend and in the course of playing around he inserted his penis inside me. He didn't force himself on me or coerce me into doing something I didn't want to do. I really didn't know what to think of it at the time. I didn't have a clue what sex was; not a clue. I had never masturbated prior to that event. I don't think he, himself, knew either. I think that all he knew was that he was doing something that he enjoyed and derived pleasure from. The first time he had an orgasm was inside me."

The night at Zippers ends at 2:00 a.m. the next morning. David and Daniel return to David's house where he had lived with his former wife. The two fool around.

Ultimately, David and Daniel split up. Today, Daniel as Danielle is in the business of offering diversity training to large corporations As of 2015 Daniel as Danielle has presented his course in diversity training to a small number of large corporations. As of 2015 Daniel as Danielle has been in the business of diversity training for over approximately fifteen years. The website that advertises his professional services in diversity training boasts very few corporate engagements. In addition to diversity training Daniel as Danielle also briefly worked for the Sherbourne Health Clinic as a research consultant. The position paid Daniel a minimum wage or thereabouts to survey people in the LGBT community to better understand the community's health needs. In addition to the diversity education services and his work as a research consultant, Daniel as Danielle volunteers his time to help to needy and mentally challenged. In 2011 the LGBT community where Daniel lives nominated him for an award to recognize his service to the local community. In 2011, however, the LGBT community ultimately gave the award to someone else.

* * * * *

It is 11:00 PM on Saturday, January 31, 2004. Six members of the transsexual community, who largely know each other via the collaboration services on a site, known to accommodate a trans audience, meet socially at a restaurant situated in Little Italy of Toronto, Ontario. Male-to-female transsexual Peter, a.k.a., Petra is also in attendance.

Peter (not his real name) is not new to what are called 'network' events. However, Peter has attended more of these events lately, than in 2003, dramatically so. Peter, who is otherwise a 48-year-old male, is scheduled to travel to Thailand in April, 2004 (only three months away), where he is to undergo a gender reassignment surgical procedure with Dr. Suporn. Peter's decision to undergo a GRS procedure is sudden. Peter reported that he only recently began to go to work as Petra, in lieu of working as Peter. In fact Peter's energy levels are otherwise quite high in general.

This evening Peter talks about his gender transition—in great detail. He talks about his electrolysis; he talks about the physical changes he observes from undergoing hormone replacement therapy; he talks about his posture; he talks about his hair, which is beautiful this evening I might add. He talks about the upcoming surgery that is scheduled with Dr. Suporn. In fact this evening Peter reports that he is observing dramatic changes to his body. Peter reports his body hair is falling out. He reports that he shaves his legs less often. He reports that he 'likes' that. Peter reports that his skin is getting softer and more pale. Peter reports that he 'likes' that too. In fact all other topics around the table at the College Street Bar where the transsexuals are seated pale in comparison to Peter's. Apparently, Peter's life as it leads up to Peter's surgery date has never been better. Peter is beaming and beautiful.

Peter's odyssey into transsexualism reportedly began on the collaboration services at the Take a Walk on the Wildside website, in December, 2000. There he interactively met a number of other self-identifying transsexuals, including Daniel, a.k.a., Danielle and Mitch a.k.a., Karen. Daniel and Mitch were both early into their respective gender transitions. Peter, Daniel and Mitch went on to become close friends. That month Peter ventured out as 'Petra Wilde'—his alter ego—as a cross-dresser and socialized with other cross-dressers at the Take a Walk on the Wildside Fetish Boutique. There Peter reportedly had an extramarital affair with Patricia Aldridge—the proprietor. Peter is otherwise a straight male although during sex with Aldridge, Peter was reportedly dressed in women's clothing, engaging in cross-gender role play.

Suddenly, disaster struck. An anonymous individual reported Peter's recent outings cross-dressed to his wife. His wife was reportedly shocked to learn that her husband of more than 18 years had such interests. Peter and his wife were parents to two children, ages 12 and 9. Peter's wife set down simple conditions for Peter to continue in the marriage—Peter had to walk away from the transvestic lifestyle. Peter at first acquiesced. However, his urges to join the transsexual lifestyle were irrepressible. Ultimately, he could not honor the agreement he made with his wife. After a short respite from the site, wildside.org, he was clandestinely back online as Petra.

In February, 2001 Peter's wife retained a lawyer and filed a petition for divorce. She forced him to move out of the matrimonial house where he had lived for years. She also effectively 'outted' him, reporting to the children that their father was in fact a cross-dresser—a fact he was not aware for years later, when he broke the news to them on his terms. The divorce was humiliating and painful. Peter was devastated by his wife's rejection, but not so devastated to walk away from the transsexual lifestyle—that triggered the divorce. In fact throughout the divorce proceeding, which was swift, Peter blamed his wife for the divorce in part. At social outings he presented as Peter and not Petra to keep a low profile and that, too, was painful. He also grew out his facial hair during that time.

Peter moved into a small apartment on Eglinton Avenue in Scarborough, Ontario—a suburb of Toronto, Ontario— that was close to work. He felt unfulfilled with cross-dressing and desperately wanted to live full time as a transsexual woman.

"[...] it soon became very obvious that life really sucked having a double identity. It's like my two half-lives just don't add up to one. So I just started taking steps – tentative steps. What can I do to make my life easier here? Well I can eradicate my facial hair. So I started laser electrolysis—to remove my facial hair. Various other things like letting my hair grow longer and learning about cosmetics—putting a wardrobe together. Eventually I said, 'ok let's see what's going on here mentally.' So I started going to see a therapist."

Peter broke the dramatic news to his family that he would undergo a complete gender transition and that he would legally become a female. His family—which included his mother and sister—had concerns, particularly in light of Peter's parental responsibilities.

"I explained it to my mom and my sister and their only concern was: 'Are you sure?' I assured them that I was, then they said, 'well if it's something you have to do then we'll support you all we can.' The kids were a little less supportive in that they weren't keen on it. They didn't try to argue with me or anything. They are fairly young and their attitudes towards me haven't changed really so it hasn't adversely affected that. It's tough for them; they're losing a father figure." Peter says.

Peter reportedly worked as an engineer for product safety testing company in Scarborough, Ontario. He effectively transitioned on the job. In 2003, Peter held closed door discussions with the Human Resources department at the company and the two parties agreed that Peter would cease to work as Peter and he would instead work as Petra. In fact in the meeting with the human resources manager, Peter reported that her seeing him as Petra was underwhelming.

Technically, at the time Peter applied to undergo the GRS procedure from Suporn, he had not yet completed a Real Life Test (RLT), and would have otherwise not qualified for the surgery had it been performed in North America. The Real Life Test requires candidates for

GRS procedures to function as a member of the opposite sex for at least one year. However, Peter arranged for the surgery in Thailand—where the requirements are considerably more lax.

Suporn requires his patients to live in the gender role of the opposite sex only six months as evidenced by two original pieces of documents. In North America the minimum time is one year. Suporn requires one referral letter from a person knowledgeable of the Gender Identity Disorder subject matter, which can include a doctor, psychiatrist, clinical psychologist, endocrinologist, or psychiatric social worker. In North America the Standards of Care document requires two referral letters; one from a qualified psychiatrist and the other from a qualified psychologist. The letter from the psychiatrist is the more important of the two. The psychiatrist must be well respected in the Gender Identity Disorder subject matter.

Peter reported that he really did not feel compelled to undergo a GRS procedure, but rationalized that undergoing one made things 'easier'. He said that if he were to ever be incarcerated, after undergoing a GRS procedure, he would never have to worry about being locked up with male inmates. Peter presumably had a concern for his safety during an incarceration. In fact, pre-operative transsexuals with such concerns can opt for protective custody. In any event Peter reportedly does not have a criminal record. Peter also said that he wanted his driver's license to match his appearance, which is that of a female. In fact, males can affect a change to the gender code on their drivers license by demonstrating that they have undergone "transsexual" surgery, but not necessarily a GRS procedure. A trachel shave qualifies as "transsexual" surgery. A Rhinoplasty procedure also qualifies as "transsexual surgery".

In April, 2003, Peter went off to Thailand. He underwent the procedure with Dr. Suporn that he so eagerly sought. In fact prior to the scheduled surgery he took 'vacation' pictures of the elephants.

One month later, Peter returned to Toronto, Ontraio as 'she', complete with a neo vagina where male genitalia had existed prior. He took time off work—lots of time—to recover. Eight weeks after the surgery, he reported that as he sat down his genital region had the sensation akin to being hit by a baseball bat, but that the unusual sensation was receding. He followed Suporn's post-surgical care instructions to the letter. He applied stents of various sizes to his neo vagina at regular intervals. Peter reported that he had no regrets undergoing the radical surgery and that if he had the choice to make over again, he would undergo the surgery again in a "heart beat".

As he reported back to work at the product safety testing company he received a warm reception. Everything appeared to be a story book ending, save the divorce from his wife.

"After I got back where I work [there were] about 150 – 170 people in the building. And there were four people who came up to me and told me that they either had friends or relatives who were 'trans'. And one told me that her best friend was 'trans'. Another one said a good friend of hers was 'trans'. One who said that a family member was 'trans'. And another who said a good friend was 'trans'." Peter says.

In fact the company where Peter worked published a picture of him on his last day as Peter in a company newsletter. The picture included a short article.

However, after the surgery, Peter's interest in networking events with large groups of transsexuals waned. Though he no longer attended large social functions, he retained his close personal relationships with Daniel and Mitch, but only for a few years.

Daniel, who underwent a GRS procedure prior to Peter, has a rather dim view of the effectiveness of GRS procedure towards gender dysphoria. Daniel divides post-operative transsexuals as ones who "survive well" and ones, who "barely survive." Daniel, reported in 2005 (one year after Peter underwent the GRS procedure with Suporn) that Peter is "on" anti-depressants where he was not prior to the surgery and that he is one of the post-operative transsexuals Daniel knows, who is "barely surviving."

Despite Daniel's bleak outlook of Peter's life, Peter is perhaps surviving better than Daniel gives him credit for. Peter as Petra reportedly has a new woman in his life, who introduced him to a new interest outside gender matters—motorcycles. Peter also retained his position at the company he originally worked at. Peter also bought a condominium where he and his girlfriend live together. Peter also drifted away from the people, who he was close to throughout his gender transition, including Daniel and Mitch.

The evening at the College Street Bar ends at 1:00 a.m. Sunday, February 1, 2004. Peter goes back to his modest apartment. Peter lies down in bed, thinking of the wonderful time he had with the group and his gender transition.

* * * * *

Mitch, a.k.a., 'Karen', reportedly began to affect a gender transition in 1999, approximately at the same time as Peter and Daniel. In fact Mitch (not his real name) vividly remembers the day he made the decision.

"I had a strong epiphany that something had to change on September 20, 1999 while I was watching the 1999 movie, Boys Don't Cry. I felt a tremendous sense of living a lie. I didn't know what route my life would take; part time; full time but I had to find out. I was 36-years-old and was tired of living a lie." Mitch says.

The only thing Mitch questioned was his resolve.

In 1996—three years prior to watching the movie, Boys Don't Cry—Mitch reported that he had broken off an intense relationship with a woman. From that point forward as he became involved with women, he disclosed that he had this indescribable feeling that he was meant to be a member of the opposite sex. He was 33-years-old at the time.

"It was a very dark period of my life and I felt that everything was up for grabs. I was so unhappy. Thoughts of transitioning had really come to the forefront of my mind in the final couple of months of my relationship (we broke up for entirely unrelated reasons) and I seriously questioned what I wanted to do. I gave a lot of thought to transitioning but I was so emotionally drained that I really felt that I needed to take some time and work though my feelings and emotions before I could really take a look at who I was and what I wanted to do." Mitch says

"Six months, maybe a year passed and I looked at my life and I looked at what transition would mean to me. I also had a lot of very frank discussions with my best friend about it. She didn't like the idea and had some very good reasons for her feelings. I also didn't like what saw. I envisioned losing my friends, my family, my job... Basically losing everyone and everything that was important to me. And the emotional upheaval! I envisioned not being able to earn a living, having to scramble to support myself. I thought that I would ultimately end up a very unhappy person. In the end, I made the best decision that I could at that time and decided not to transition. However, I also resolved that I would be totally up front about who I was as a person with anyone that I was in a relationship with. I was going to be totally open about who I was as a person and that although I'd considered transitioning that I had decided against it."

Although, as he dropped the bomb on the women who came into his life that he had these thoughts, none reportedly ended their relationship with him.

"I think that I was very open and honest with them and in the end, they respected that above all and worked together to deal with issues. I helped a number of them to come to terms with their own gender identity." Mitch says.

When Mitch made the decision to transition, he joined Xpressions—a social club for cross-dressers. However, his tenure with Xpressions was short lived. He desperately sought people, who were ahead of him and not behind. He met 'Daniel', a.k.a., 'Danielle'. In fact Mitch credits Daniel with kick starting his transition.

"[Daniel] was actually the one who threw me off the diving board and told me to swim when I was dithering about when to go full time."

Like Daniel, Mitch spent exorbitant time on the Internet, consuming material on transsexualism.

"I read literally hundreds of personal pages as well as spending quite a bit of time on the chat service hosted by the Transgendered Network International (e.g. www.tgni.com). I learned more about self-respect, self-worth, transition, surgery, the nuts and bolts [of transsexualism] from that chat room than any other resource." Mitch says.

In 1999, Mitch worked for Canadian Imperial Bank of Commerce (CIBC). At CIBC he worked on the trading floor with three hundred other people. His responsibilities involved affecting currency transactions. On any given day Mitch booked an average of $4,000,000,000 worth of transactions. Although he was not in a sales capacity, he earned a decent salary, flirting with the $100,000 mark. He was reportedly a man of importance and highly respected.

Mitch lived in a 900 square foot condominium on Bay Street—that he owned outright. The unit faced one side of the downtown Toronto skyline. From the unit at night the view is spectacular. His commute to work was brief and involved three subway stops.

On Daniel's encouragement, Mitch contacted the Human Resources department to feel out the company's policy on a change of sex.

"I started by exchanging a number of emails with people in Human Resources on an anonymous basis. In those emails, I was basically asking them what their policy was towards transsexualism, what should I expect job-wise, career-wise." Mitch says.

With his heart pounding Mitch anxiously anticipated each response. The e-mails back were brief, asking Mitch to identify himself so the matter could be discussed directly.

Ultimately, on July 4, 2001 Mitch met with a woman from the HR department at the food court at Scotia Plaza. Mitch regarded the date as his independence day. Mitch was smartly cross-dressed as he met with the woman from HR. Mitch reported that the woman, whom he met with, was open minded and interested to learn more about Mitch's transsexualism. In fact, the biggest issue reportedly revolved around the use of the woman's washroom.

There were several other meetings; one involved meeting with the Vice President of his department.

As the bank at that time didn't have any policies or procedures dealing with the gender transitions of its employees, the group at the meeting struggled with what to do. The group discussed bringing in a councilor and scheduling meetings involving small groups of people, 15 or so, where people could be given a brief background of Mitch's transition, but that idea was abandoned. Mitch's transition and how to deal with it went as high as the Bank's Board of Directors. At one point in the meeting the Vice President reportedly was of the opinion that the matter was no big deal and that's the way it should dealt with.

"They pulled my direct department, about 15 people into a meeting with my Senior Business Line Manager and VP from HR chairing I was specifically not in attendance. This was their time. I was not in attendance. She basically told everyone that I was transsexual and would be transitioning at work in about a week and a half. She then discussed a lot of the things that I'd been educating her about over the previous month or so as well as the Bank's stance regarding it, which was that it was a personal issue. She also made it very clear that I enjoyed the absolute, unconditional support of upper management, including the Board of Directors and that there was a zero tolerance policy towards any kind of harassment, discrimination of me no matter how minute and that violating that policy was grounds for immediate dismissal." Mitch says.

"My second line manager said he was blown away by people's reactions. No one raised an issue or had a comment. They were all supportive, concerned and hoped my transition would proceed without a hitch."

Mitch took a week off and when he returned, he returned as Karen and not Mitch. The day of his return was uneventful and Mitch was ecstatic.

"It was the best possible outcome. It was just another day, coming back from vacation and getting back in the swing. It started off with a really nice catered breakfast in one of the Board Rooms before work where my co-workers could come and meet the "new" me if they wanted and then it was time to go down to the Floor. Most simply came up to congratulate me, some were curious about what I was wearing, and others were simply curious to see what I looked like as a woman." Mitch says.

As time progressed all of Mitch's prior anxieties flat-lined. He was reportedly whole, but as whole as he was, he continued to explore and think about the meaning of gender, and what gender meant in a rich social context.

"Three months have past and I'm just another woman on the floor. I come in, I chit-chat with the other women, and I'm accepted. How could I ask for anything more? I go out to lunch with friends, we socialize. I couldn't have these kinds of relationships as 'Mitch'."

After successfully affecting a gender transition at CIBC, Mitch turned his attention to undergoing a GRS procedure. He carefully investigated the work of the various plastic surgeons, who offer surgical procedures to transsexual patients. In fact, Mitch traversed North America, interviewing surgeon after surgeon with regard to their work and what they could offer him. He sat down with Dr. Pierre Brassard in Montreal, Quebec, Dr. Toby Meltzer in Scottsdale, Arizona and Dr. Eugene Schrang in Neeham, Wisconsin. Of all the surgeons he met, he was least impressed with Brassard. Mitch characterized Brassard as somewhat passive and non-animated and that he had to draw Brassard out to get any information from him at all. He also looked into Dr. Suporn, in Thailand, but discounted him. Mitch ultimately decided on Schrang on the strength of the cosmetic result.

In January, 2002, CIBC reorganized Mitch's department. Half of his department split off to continue to affect currency transactions and the other half formed a risk management group. The risk management group oversaw the work of the group that affected currency transactions and that of a number of other groups. Mitch, who was given the choice of where to go, opted to help form the risk management group. By this time Mitch had 12 years of experience at CIBC. The change was effectively a lateral move and a good fit for his skill set. Further, CIBC continued Mitch's grade and salary at the same level. While working for the risk management group, he prepared financial reports for senior executives who relied on the facts Mitch put together to affect business decisions, involving hundreds of billions of dollars.

In October, 2002, Mitch reportedly took time off work and underwent a GRS procedure with Dr. Eugene Schrang in Neehan, Wisconsin. At that time the Canadian dollar hit an all time low against the green back and the surgery cost Mitch approximately CDN$30,000.00. Unlike Daniel however, Mitch was pleased with the result, and felt he got his money's worth. He reported that he had some bleeding from the surgery, but that the bleeding was minor compared to what he heard other people had to deal with. In total he took approximately two months off work to convalesce.

When he returned to work as 'she' he continued to prepare financial reports. However, to supplement the day-to-day grind of preparing reports, he volunteered his time at the personnel resources group within the bank. The CIBC personnel resources group is a volunteer run organization within the bank that helps minorities and homosexuals cope with oppression. While working for the personnel resource group Mitch became very involved in bank-sponsored, gay, lesbian, bisexual transsexual (GLBT) events. He helped organize marches and oversee the construction of CIBC Toronto Pride Parade float. In fact Mitch won the 2007 CIBC Equity & Diversity Initiatives for Visionary Leadership Award (e.g., the CIBC Diversity Award for short). CIBC limits the award to a handful of recipients each year. Further, on the departure of the co-chair of CIBC's Pride Group, Mitch was appointed to the acting co-chair of CIBC's Pride Group, while the bank looked to back fill the position.

Eventually, the volunteer work coupled with the "number crunching" was reportedly no longer fulfilling at CIBC and in May, 2007, after eighteen years of service with the bank, Mitch handed in his resignation. That year he had a lead for a job as a National Director of Diversity and Inclusion for a large company, however, the job didn't materialize. Instead, in November, 2008, Mitch established a sole proprietorship to offer diversity awareness training services. Mitch reportedly has two partners, who are involved in the business, and has had some discussions with yet a third individual about handling seminars on creating workspaces for people with disabilities. Mitch has also made inquiries at CIBC and reported that he has been in talks with people at the bank on three separate occasions about a contract opportunity, should things not materialize in diversity awareness training services. In any event the lack of earning an income over 2008 has reportedly not fazed him.

Mitch admitted that developing intimate relationships as a male-to-female transsexual has had its challenges.

"When I volunteer that I am a 'trans' woman, I get one of three reactions. Either the woman thinks transsexuality is 'out there' and it's fucked up, or transsexuality and my ability to overcome transsexuality speaks positively about my character or transsexuality really doesn't matter." Mitch says.

Mitch's relationships with natal women reportedly last from as short as one month to as long as 18 months—not very different that that of many men, seeking intimate relationships with heterosexual women.

Again, Daniel, who underwent a GRS procedure prior to Mitch, has a rather dim view of the effectiveness of GRS procedure towards gender dysphoria. Again, Daniel divides post-operative transsexuals as ones who "survive well" and ones, who "barely survive." Daniel, reported in 2005 (three years after Mitch underwent a GRS procedure from Schrang) that Mitch, like Peter, is "on" anti-depressants where he was not prior to the surgery, but unlike Peter, Mitch is one of the few post-operative transsexuals who is "surviving well."

Despite Daniel's alarming critique, Mitch is perhaps surviving better than Daniel gives him credit for. Mitch is reportedly as comfortable being Karen five years after surgery as he was one year leading up to the surgery. He is reportedly passable by third party testimony, which is rare in males who affect a gender transition relatively late in life. The fact that prior to May, 2007 he retained his career at CIBC, coupled with the fact that he is able to engage in intimate relationships as Karen, coupled with the fact that he is able to develop personal relationships as Karen outside the transsexual community may be the reason behind his comfort level.

Yet, Mitch's continued focus on all things 'trans' and 'giving back to the community' is troubling. Mitch's continued interest in other people's transition stories is akin to the recovering alcoholic who tends bar. No matter how well the alcoholic is coping after rehab, the temptation to stray towards thoughts of consuming alcohol is omnipresent. In the case of the individual who has a behavioral addiction, underlying transsexualism, the temptation to stray towards thoughts of reliving one's own gender transition through hearing the transition stories of other transsexuals in the guise of diversity awareness training services and volunteer organizations like CIBC's personnel resources group is also omnipresent.

"I'm me. [I'm] just a woman with a 'trans' medical history. Now I'm just 'Karen'. Life has been good." Mitch says.

Ultimately, Mitch's employment at CIBC ended. Although Mitch tried to find a job at another financial institution he had considerable difficulty. Ultimately, Mitch as Karen abandoned the search for a job to replace the one he had at CIBC and instead opted to pursue a Master's Degree in Education. As of 2015 Mitch as Karen was still unemployed.

*　*　*　*　*

Adam was born into an English family in 1937 just prior to the outset of World War II.
He was the youngest of a family of two children. Adam (not his real name) had a sister, who
was seven years older. She was born in 1930. Adam's father served in France during the Great
War (World War I). On returning from the European theatre Adam's father suffered from Post
Traumatic Stress Disorder. Post Traumatic Stress Disorder was known as 'Shell Shock' during
the period. Adam reported that his father would become very frighten by incidental noises.
The sound from a door slamming would reportedly set off Adam's father.

At the time of his birth, Adam's family lived in Leicester in Central England, but when in
1942 the German Luftwaffe bombed the city, Adam's family moved to Coventry, the next
town West of Leicester.

Adam did not have an unusual childhood. As a child Adam played with boys, however,
he did not like participating in organized sports very much. He reported that he liked skipping
rope with girls his age.

Adam's odyssey into transsexualism began at age nine. At age nine Adam cross-dressed,
wearing his sister's clothes. Adam, who by that age knew society regarded cross-gender
behavior in males as abhorrent, cross-dressed in private.

At one point Adam's sister caught him, wearing her clothes. She told her father of
Adam's transgressions, exposing him as a cross-dresser. Adam characterized the incident as
'traumatic'.

At age nine Adam was fascinated with his sister. At the age of sixteen Adam's sister was
rapidly becoming a mature, adult woman. At one point Adam thought that a liquid in a bottle
that sat on his sister's dresser caused her breasts to grow. That one incident may have given
Adam the idea that femaleness could be medically constructed and not purely the result of
nature. At the time, Adam's family, including his parents, never discussed the difference
between the sexes nor what sex and intimacy were all about. In effect Adam had no materials
to draw from to know how males and females anatomically differ or what intimacy involved.

Adam reportedly first masturbated at age 13 at the onset of puberty. Adam reported that
while attending boarding school he witnessed another adolescent male fondle his own genitals.
He emulated him.

Adam academic performance was mixed. Although he failed England's National 11+
exams, he was one of the few students to pass England's National 13+ exams. That set him

apart. After passing England's National 13+ exams, Adam attended a more elite High School than his peers.

Adam reported an incident when one girl from his class came to class wearing cosmetics. The school policy at the time barred girls from attending class while wearing cosmetics. The school authorities directed the girl to wash off the cosmetics from her face. Adam reportedly felt jealous that although the girl was directed to wash off the cosmetics she was wearing, she was free to attend class, wearing cosmetics, while he was not.

In 1954 at age 17, while attending boarding school, Adam developed a boyfriend-girlfriend relationship with a woman of the same age as he, who attended the same school that he attended. Adam was at this time quite naïve in the ways of sexual intimacy. The extent of the sexual aspect of Adam's relationship with the woman never advanced beyond a kiss. A relationship with a woman that went beyond a kiss was beyond Adam's comprehension at the time.

In 1955 at age 18 Adam was conscripted into the British armed forces and opted to serve with the Royal Air Force (RAF). Adam was immediately stationed at an RAF base in Germany. On his departure to Germany Adam asked his then girlfriend in England to marry him. However, while engaged and while stationed at an RAF base in Germany, Adam developed a relationship with a German woman. Adam did not have sex with her either. However, on a visit to the house where the woman lived, while she was out, he put on her clothes.

Adam also cross-dressed on the RAF base in private. Having no access to women's clothing Adam fashioned a towel as a make-shift skirt. Even in these crude circumstances Adam's cross-dressing followed a predictable pattern. Adam put on the women clothing (a towel for the most part), masturbated then took off the clothing.

Adam's typical cross-dressing episode lasted ten minutes. In the case where he had access to genuine women's clothing Adam reported that he largely focused his attention on the 'feel' of the clothing.

In 1958 at age 21 Adam received an honorable discharge from the British armed forces and returned to England. On his return Adam attended College and majored in the sciences and more specifically Atomic Physics. He took Math and Physics courses. Adam's decision to closely follow in his father's footsteps was his own choosing.

"I've always been interested in science. My father was a great engineer—mechanic. I had a great interest in science and electronics." Adam says.

On his graduation Adam immediately worked for the British Defense Department. He held security clearances and worked on classified projects, pertaining to atomic weapons development. Adam did the design and development of the central core of the triggering system of a number of England's atomic weapons. He lived in a hostel at the facility.

While at work and while eating lunch, Adam met a young woman, a Chemist named Jayne. Jayne (not her real name) also worked at the facility Adam did. Jayne worked in Health Physics. She studied the effects of exposure to nuclear radiation on human subjects.

At first Adam was put off by Jayne, but offered to help her on one of her projects. He invited her to his room. One invitation led to another and Jayne was a constant fixture in Adam's room.

At one point Jayne told Adam her period had stopped and she thought she was pregnant. Adam had again proposed marriage, but he did so under the weight of what he believed were damning circumstances. Apparently, at that time in England a woman could sue for breach of promise. When Adam proposed to Jayne, Adam was intimidated by the fact that Jayne's family were Victorian (e.g., strict). In fact Adam's new fiancé was not pregnant.

"I felt I had been cheated. I felt that I had been trapped. But because I was a religious person as well, I felt committed to carry on and marry her as I said I would." Adam says.

As he promised Adam went on to marry Jayne. He went on a honeymoon with her. However, the couple did not engage in sexual intercourse at the time of the marriage, nor immediately after. In fact, the couple did not engage in sexual intercourse until one year after the marriage took place. In fact, it was Adam, who introduced intercourse to the relationship.

Adam reported that his wife had little interest in sex.

"She would always say, 'sex is overrated'". Adam says.

The couple's mutual interest in intercourse was short lived. Adam reported after he had introduced intercourse to the relationship, that he had mutually gratuitous intercourse with his wife for a period of nine months.

As much as Adam hated the idea of being married, he saw the opportunity to enhance his cross-dressing experiences. He again had access to women's clothing.

During the early years of his marriage, Adam continued to cross-dress. Adam cross-dressed and masturbated to the feel of his wife's clothing approximately once per week. He waited for his wife to be out-of-the-house to masturbate, while wearing her clothing.

Adam, also, began to introduce female under garments to his wardrobe. He wore women's underwear to work.

"I would wear it and throw it away." Adam says.

In 1964 Adam and his wife became parents. In 1964 Adam and his wife had their first child—a girl. A year later, in 1965 Adam and his wife had their second child—another girl. Seven year later, Adam urged his wife to have a third child, hoping for a boy. Seven year later, in 1972 Adam and his wife had their third child and again it was a girl. One year later, in 1973 Adam and his wife had their fourth child—a boy.

After the birth of his first child, in 1964 Adam also began to experiment outside of the house. Adam attributes his experimentation to his scientific background.

"I would test myself I would go out and see what was acceptable. I would go out in female clothing." Adam says.

In 1964 at age twenty-seven Adam's public cross-dressing episodes were limited to car trips cross-dressed and bike rides cross-dressed, where the risk of being discovered was low. It wasn't satisfying. Adam wanted a more socially rewarding experience cross-dressed. Eventually, he found it.

In 1968 at age thirty-one Adam connected with other men like himself. He found an advertisement for a club that catered to men, who cross-dressed, then as he researched he found more advertisements. One was for the Beaumont club.

"I joined various clubs. We used to meet at people's houses and then we used to go to have Christmas dinners together at a restaurant which was especially arranged for us to go to [these] events." Adam says.

Adam also saw a woman, a registered nurse, who provided a facility to serve men, largely heterosexual men, who wanted to look like women and engage in social activities. The female nurse not only organized outings, but also she invited members of the media to interview men cross-dressed in public. A member of the media interviewed Adam.

By 1974 Adam earned 16,000 British pounds (the equivalent of 25,000 US dollars per annum). In 1974 Adam and his wife and their family of four children lived a middle class existence. They owned in a large house in an affluent neighborhood, two cars and a boat.

Adam came out to his wife at the time he married her. He told her he liked to wear women's clothing. It is not clear how Adam's wife reacted to the news or whether his wife truly understood the ramifications of being married to a cross-dresser. However, Adam never exposed his proclivities towards cross-dressing—his cross-dressing lifestyle—to his wife. His wife was uninvolved in his extracurricular affairs. If his wife spoke to Adam while knowing he was cross-dressed, she did so behind closed doors, where she didn't have to see him first hand.

By 1980 Adam's masturbation thoughts had dramatically changed. Where originally Adam masturbated to the thought of women's clothing and the 'feel' of women's clothing in 1980 Adam began to masturbate to the fantasy of being female and in particular the fantasy of being a female fashion model.

In 1980 Adam began to live out his fantasies. Adam began to engage in what he regarded as women's activities, including fabricating his own women's clothing and modeling women's clothing. As he began to engage in women's activities, as he began to model women's clothing, his proclivities towards cross-dressing not only dramatically increased and he also began to consider cross-living.

"I am on stage and I'm walking down a cat walk with clothes I have designed myself and I get top marks from other girls, I had then felt I had achieved something. I even paraded on the cat walk with a bathing costume even though I hadn't had surgery at the time." Adam says.

In 1980 Adam also saw a doctor to obtain a prescription for synthetic estrogens and began hormone replacement therapy.

In 1980 Adam no longer worked for the Defense Department, but instead worked for an electronics company.

Over the next six years Adam kept his gender transition a secret from his co-workers. Adam had the benefit of having a private office, which afforded him the ability to lock the door as he cross-dressed. He thought nothing of wearing male appropriate clothing from the waist up, while wearing female appropriate clothing from the waist down. When he engaged co-workers while seated at his desk, no one was the wiser, or so he thought.

Adam reported that his productivity declined over the course of his gender transition. He was passed up for promotions that other workers received. Adam reported that his thoughts were 'elsewhere'.

"I wasn't concentrating on my work. I was concentrating on my transition." Adam says.

In effect Adam kept up the appearance at work that nothing was out-of-the-ordinary.

Adam also kept his gender transition a secret from his family. His family members, including his wife, were completely unaware of his gender transition.

In 1986 Adam attended his daughter's wedding. He presented himself as a man and wore male appropriate clothing. He felt horrible doing so.

In 1987 the electronics firm that Adam worked for down-sized as a result of defense cuts and Adam's job was deemed redundant. Effectively, Adam was laid-off. In 1987 after twenty-eight years of marriage Adam and his wife also divorced.

After Adam was laid off in 1987, after his divorced that same year, Adam switched careers. He became an architect. He was then fifty-years-old and self-employed. Adam then worked as Evelyn and presented as a female to his clients. He reportedly passed well and his clients were unaware that he was a male-to-female transsexual.

Adam as Evelyn also became a professional dancer.

By this time his wife had also switched careers. She was working as a public school teacher.

Between 1980 and 1987, while Adam was undergoing HRT, Adam's ability to get erect dramatically diminished. Adam's sex life with his wife, which was non-existent, also got worse. The couple rarely had sex.

In 1985 Adam attended the main NHS Gender Identity Clinic at Charing Cross hospital, west London as a candidate to receive a publicly funded GRS procedure. Dr. Russell Reid attended to his case. Reid (or was at the time) is a renowned British psychiatrist, specializing in sexual and gender-related conditions. Reid is particularly known for his work with patients exhibiting gender identity disorder.

In retrospect Adam believes he was misdiagnosed while receiving care at the GIC and put on a fast track to undergo a GRS procedure.

"I think I was badly advised, because [the GIC] put me with more estrogens and they were handed out like sweets." Adam says.

In fact Adam had not lived in the female role at the time he was undergoing HRT under the care of the doctors at the GIC. Adam had lied about cross-living, fearing the doctors would discontinue the hormone replacement therapy if he told them the truth about how he was living at the time.

"My advisor said that if I didn't change, I would probably commit suicide." Adam says.

"They persuaded me to go through with it."

After his divorce in 1987 from Jayne Adam met a woman, who (unfortunately for Adam) had sexual interest in male-to-female transsexuals. Her name was Ruth. Ruth (not her real name) took an interest in Adam's gender transition. Ruth found Adam attractive as Evelyn. Adam liked Ruth. Adam thought Ruth, too, was hot, but more so he liked the fact that she found him attractive as a male-to-female transsexual. In effect Ruth's interest in Adam as a transsexual woman corroborated with his interests in cross-living, further propelling him into the decision to undergo the surgery.

"I believed I was being tough and manly going through with this thing to impress her with the intent of escaping at the last moment." Adam says.

Unfortunately, Adam didn't escape. He didn't cancel at the last minute. On April 30, 1989 Adam underwent a publicly funded GRS procedure. Dr. James Dalrymple performed the surgery on Adam. Psychiatrist Dr. Russell Reid of the GIC wrote Adam's letter recommending the GRS procedure in his case.

Two years after the surgery, in 1991 Adam wrote Dr. Reid a letter, stating that he didn't like what had happened to him and that he regretted undergoing the GRS procedure. In his letter Adam also made inquiries about what could be done to reverse the procedure.

On March 4, 1991 Dr. Reid responded. In his letter to Adam the psychiatrist told Adam that he thought it was in Adam's best interest to make an effort to adjust to life regardless of his gender role. [2]

Adam tried to live out his life as Evelyn, but felt awkward doing so. In fact at one point Adam as Evelyn 'cross-dressed' as a man. Further, Adam failed to regularly apply stents to maintain his neo vagina. He believes that his neo vagina is now fully closed.

Sadly, none of Adam's four children maintain a relationship with him.

"[my oldest daughter said to me], 'no one wants to know you anymore.'" Adam says.

Adam attributes the rift in his relationship with his children to his behavioral addiction towards transitioning.

"I am appalled at myself for what happened at the time." Adam says.

"I wanted to improve [my presentation] all the time to get more. This is why I explored these various clubs. This is why I became a [professional] dancer. This is why I became a fashion model. This is why I took up singing. I was doing as much as I could. I tried to assimilate."

Adam reported that he never achieved orgasm after undergoing the GRS procedure, but was able to prior to the surgery.

In 2006 the General Medical Council (GMC), the regulatory body of doctors in the UK, investigated Dr. Russell Reid, following the complaints brought by four doctors of the main NHS Gender Identity Clinic at Charing Cross hospital, west London, and a number of his patients. The complaint alleged that Reid breached the international standards of care, set by the Harry Benjamin International Gender Dysphoria Association (HBIGDA) by inappropriately prescribing sex hormones to patients and referring them to undergo a gender reassignment surgical procedure without adequate assessment.

On May 25, 2007 a GMC fitness to practice panel ordered Dr. Russell Reid to adhere to strict conditions in prescribing sex hormones to patients over a twelve month period, effectively stripping Reid of his autonomy to direct the gender transitions of transsexual candidates. [3]

* * * * *

Blanchard of the CAMH has developed an objective test to determine the presence of autogynephilia. Michael Bailey also developed an objective test for autogynephilia. [4] However, Bailey's test is much simpler than Blanchard's, which reportedly includes approximately one hundred questions.

In fact only one question is necessary to determine autogynephilia in males.

> "Has anyone who knew you as a toddler at 24 months or thereabouts reported that you tugged on your clothing in an apparent attempt to remove gender appropriate clothing?"

If yes, the person suffers from a gender identity disorder. If in the case of a male child the answer is no and if the first incidence of proclivities towards cross gender behavior occurs at age 5 or thereabouts or age 11 or thereabouts, the male-to-female transsexual suffers from an affected autogynephilic disorder. If in the case of a female child the answer is no and if the first incidence of proclivities towards cross gender behavior occurs at age 4 or thereabouts or age 10 or thereabouts, the female-to-male transsexual suffers from an affected autoandrophilic disorder.

The reason stems from the fact that gender identity is formed early in life, within the first 12 to 18 months of age, earlier than arousal-based, sexual identity (e.g., what sexually arouses you).

Auto-erotic addiction, e.g., autogynephilia or in the case of a female autoandrophilia is not necessarily masturbatory. This is particularly true in the latter stages of the addiction. Yet, all transsexuals share the common experience of masturbating to the thought of turning into a member of the opposite sex as they sexually develop, which typically begins at age 5 or age 11. Gender dysphoria does not occur at the same time as sexual identity, which occurs at age 14 months or thereabouts. This is an important distinction. David Reimer, who was biologically male though raised as a girl under the direction of psychologist John Money, exhibited gender dysphoria at 14 months of age. After transition back to living as male Reimer never spoke of life as a member of the opposite in a gratuitous manner. Reimer regarded the experience as traumatic.

Auto-erotic addiction, e.g., autogynephilia or in the case of a female autoandrophilia is easy to detect.

In the case of a male it is all about him as 'her'. The autogynephile talks endlessly about his transition, his electrolysis sessions, the development of his breasts, the transformation of his body, his medication protocol. The autogynephile can't talk enough about his gender transition—which was barely on the radar screen a short time prior to transition—even after undergoing procedure after procedure, nor consume enough of other people's gender transition stories. The autogynephile has volumes of images and videos himself in all stages of transition, with particular emphasis on the images in which he looks remotely 'passable' as a member of the opposite sex.

The autogynephile demands the people around him corroborate with his living, breathing delusion that he is female. The presence of his otherwise healthy sexual anatomy is in his mind a birth defect due to some nebulous accident even though so-called gender dysphoria is not present in any other species in the animal kingdom.

The autogynephile demands attention and sympathy from the greater population. The autogynephile regards himself as somewhat passable even though to the greater population, he clearly isn't and often becomes outraged at any violation to his fragile ego or mis-gendering

"Even the most passable transsexuals fail to pass 100% of the time," Maxine Peterson says. Maxine Peterson, a.k.a., Leonard Clemensen, is a clinical researcher, who had worked alongside psychologist Dr. Ray Blanchard at the Gender Identity Clinic at the CAMH prior to his retirement.

The autogynephile gravitates to careers, organizations and/or groups that bring gender transition omnipresent in his life, in a similar manner to a pedophile who gravitates to careers that bring him close to children as either a camp counselor or an elementary school teacher. The autogynephile rarely breaks away from his gender transition. When he does, he re-lives his gender transition vicariously through the gender transitions of others. He rationalizes this practice as 'giving back to the community' or diversity training or writing a book on the transsexual subject matter or similar activity.

The autogynephile objects to the staff at first wave centralized gender identity clinics—who know how to detect autogynephilia in males—for being insensitive to his indisputable need to undergo a GRS procedure. However, when these first wave centralized gender identity clinics offer the autogynephile non-invasive treatments to GRS procedures, the autogynephile shows no interest. The autogynephile will opt for treatments and therapies that corroborate with his self-image, e.g., physical changes to his body to imitate either a member of the opposite sex or the plasticized equivalent. The autogynephile relies entirely on self-diagnosis and the diagnosis of primary care physicians of third-wave health care providers and rogue general practitioners, who possess little working experience with real post-operative transsexuals and their real life health challenges.

The autogynephile goes where the waters are warm and the music resonates to his ears.

* * * * *

Transsexualism as a phenomenon meets all the diagnostic characteristics of a behavioral addiction. (See Appendix D—Transsexualism & the Diagnostic Characteristics of Behavioral Addictions)

Auto eroticism, e.g., autogynephilia in males, e.g. autoandrophilia in females, e.g., transsexualism, works in the same manner as sex addiction, Internet addiction or gambling addiction, and is caused by a faulty reward system in the human brain. Dr. Ray Blanchard of the CAMH coined the term, autogynephilia, to denote a male's paraphilic tendency to be sexually aroused by the thought or image of himself as a woman.

*　*　*　*　*

Auto erotic addiction, e.g., autogynephilic addiction in males, e.g. autoandrophilic addiction in females, e.g., transsexualism as a disorder has a predictable lifecycle, including five distinct phases. Each phase is marked by the person's sex drive and proximity to a major invasive surgical procedure, affecting the testosterone level, like a GRS procedure or similar procedure that protracts the testes.

Phase 1, *Contraction*, of the disorder begins as the person sexually develops, either at age 5 or at age 11 or thereabouts. At age 5, the toddler already has a gender identity that is immutable. Blanchard believes autogynephilia sets in due to a failure in the male child to fully externalize the female love object. In any event the autogynephilic seeds are firmly planted. The child will either develop to be a cross-dresser or transsexual or she-male or a man with sexual interest in transvestites, transsexuals or she-males or some combination as sexual disorders rarely operate in isolation.

Phase 2, *Affirmation*, of the disorder begins as the person seizes on the consideration of a gender transition, involving a GRS procedure or similar feminizing procedure where the testes are protracted. (The phase may also begin when the person simply commits to undergo hormone replacement therapy. This is referred to a partial autogynephilia.) A person suffering from an affected autogynephilic addiction disorder–phase 2 will begin to experience obsessive compulsive thoughts. A number of these thoughts may revolve around a GRS procedure or similar feminizing procedure. The person may in fact have a history of obsessive compulsive thoughts due to other ancillary mental disorders. The compulsive obsessive thoughts, themselves, may or may not cause sexual arousal. However, the person, who has an affected autogynephilic disorder, can revisit autogynephilic thoughts at all phases of the disorder to bring about sexual arousal. In phase 2, autogynephilia moves beyond a sexual disorder and takes the characteristics of a behavioral addiction. [5] (See Appendix D—Transsexualism & the Diagnostic Characteristics of Behavioral Addictions).

Phase 3, *Escalation*, of the disorder sets in during the period leading up to a major invasive feminizing procedure, affecting the testosterone level, like a GRS procedure. By phase 3 the disorder firmly takes on the characteristics of a behavioral addiction.

A person, who suffers from an autogynephilic addiction–phase 3, produces high amounts of endorphins in the pituitary gland of the brain that in turn causes the release of high levels of dopamine—the chemical in the brain that affects the reward system—in the synapses. In fact the pituitary gland in the human brain typically produces endorphins during orgasm or similar sexual stimulus. The consideration of a gender transition qualifies as a sexual stimulus.

As dopamine is the gasoline of the reward system, the brain's own inhibitory system acts as the brake, stabilizing mood. When the person's inhibitory system, which is attenuated by the chemical, gamma aminobutyric acid (GABA), fails to function correctly, the brains never feels a sense of satisfaction. Thus, the endorphins of the person, who has an affected autogynephilic addiction disorder–phase 3, acts on the GABA as well, suppressing it. In effect the endorphins shut down the GABA, causing a fault in this critical function.

As the brain rewards the practice of autogynephilia it does so like any other behavioral addiction. The unusually high levels of dopamine coupled with the throttling of the GABA cause euphoria. A person, who suffers from an affected autogynephilic addiction disorder–phase 3, will typically report some degree of euphoria. In fact as the person takes steps in their respective gender transition, each one seems the correct one.

A person, who suffers from an affected autogynephilic addiction disorder–phase 3 and who also possesses a sex drive, will invoke masturbation thoughts that are not indigenous to a member of the opposite sex. For example, natal females do not masturbate to the thought of developing breasts and switching sexes or similar rituals of becoming female. In fact natal females are not interested in transsexual topics at all. Yet, typically the afflicted individual is utterly incapable of recognizing his distinct sexual practices are not of the norm. In the case where the afflicted individual is, he evades the issue. In the absence of a sex drive, the person, who finds it exciting to live as a member of the opposite sex, is still producing endorphins and in turn the endorphins continue to act on the synapse in the same manner.

A person, who has an affected autogynephilic addiction disorder–phase 3 may report his life 'improving'. He may even report that his life has never been better. Generally, a person who suffers from an affected autogynephilic addiction disorder–phase 3 is not productive at work. The person is typically incapable of focusing on matters outside of his gender transition—even the mundane.

A person who suffers from an affected autogynephilic addiction disorder–phase 3 will entertain over valued thoughts. The person may believe, for example, that their sex or gender transition doesn't really matter on the job. The person may believe that they are more attractive as a member of the opposite sex and are really meant to be a member of the opposite sex. The person may at age 40 or thereabouts, pose the question to himself, "why not spend the second half of my life as a member of the opposite sex?" which is an impossibility.

In addition to euphoria and over valued thoughts, a person who has an affected autogynephilic addiction disorder–phase 3 will experience high levels of energy. The person may believe he is more mentally acute as a member of the opposite sex. The person may also subscribe to thoughts of grandeur; for example the belief that his life story is news worthy and should be published as a book and sold nationwide. In fact the person may publish a book only to see the stock of books be unsold and written off. The person may also experience erratic mood swings and irritability. If the person is undergoing hormone replacement therapy, the synthetic estrogens will accentuate any such mood swings. If that happens, the person will begin to exercise poor judgment similar to the manic episode phase of an affected bi-polar disorder. The person may spend excessive amounts of money on wardrobe, etc. for the new person he is planning to bring into the world. In fact, the person, who has an affected autogynephilic addiction disorder–phase 3, is not bringing anyone new into the world. On the contrary, he is bringing to bear enormous and unnecessary health challenges onto his life.

Phase 4, *Rapid Cycling*, sets in after the person has fully recovered from a major invasive feminizing procedure, affecting the testosterone level, where the testes are protracted. In phase 4 of the disorder the person may begin to experience dramatic degradation in his sex drive. However, on the consideration of yet additional feminizing procedures, he may elevate his otherwise weak sex drive to a level similar to an affected autogynephilic addiction disorder–phase 3. In the process the person migrates from one gender-related obsession to another. Effectively, a person, who suffers from an affected autogynephilic addiction disorder–phase 4, finds himself in a vicious cycle, requiring more and more feminizing procedures to bring about the same degree of satisfaction.

Phase 5, *Exhaustion*, sets in after the person undergoes a number of feminizing procedures, following a major invasive feminizing procedure, like a GRS procedure or orchiectomy, where the testes are protracted and no longer exhibits a sex drive. The person may start to suffer from a number of mood related disorders, including depression. A person, suffering from an affected autogynephilia disorder–phase 5 may report that his life is in turmoil, a "wreck" or report fresh disasters. The person may attempt to commit suicide or even worse succeed at committing suicide. The vast majority of people, who suffer from an affected autogynephilic addiction disorder–phase 5, exercise gender apathy, projecting no gender expression whatsoever. In effect these people effectively revert back to living as a male, but do not advertise themselves as such. In these cases the person will not explicitly state regret; rather, they justify their gender transition as being correct and well founded. Often, it is a abject failure and they watch their respective life get worse and not better. In a very small number of cases, the person may explicitly state regret or revert back to living as a visible male. At that time, the person breaks free of the behavioral addiction, underlying the disorder, but at great cost to his compromised endocrine system. Clinical depression, which can be caused by a compromise endocrine system, is prevalent in the post-operative transsexual community.

In fact a Swedish study, published on February 22, 2011, that was conducted over a thirty year time span from 1973 to 2003 and consists of 324 post-operative transsexuals (191 male-to-female transsexuals, 133 female-to-male transsexuals) found that post-operative transsexuals were twenty times more likely to commit suicide than the general population. The study further found that post-operative transsexuals have considerably higher risks for mortality, suicidal behavior, and psychological morbidity than the greater population. The study concluded that although surgery may alleviate gender dysphoria, surgical procedures do not effectively treat transsexualism and more improved psychiatric and somatic care is required. [6]

The psychiatric implications at looking at the autogynephilic disorder as a behavioral addiction are clear. Break the cycle of behavioral addiction and you break the disorder. If you can do that, you can treat the cause of the disorder and not the symptom—the delusion that otherwise healthy sexual anatomy is a birth defect. (See Chapter 12, the Treatment Alternatives to GRS procedures).

references

The Transsexual Delusion

[1] BLANCHARD STUDY ON PREDICTING POSTOPERATIVE REGRETS

Blanchard et al. conducted a study to investigate whether heterosexual males are more likely to regret sex reassignment surgery than homosexual males or females. The study is quantitative in nature. The study found 29% of heterosexual male-to-female transsexuals explicitly expressed regret undergoing the surgical procedure. The study found no homosexual males or females to explicitly regret undergoing the surgical procedure, indicating a high correlation between sexual orientation and the propensity to regret undergoing the surgical procedure.

Prediction of Regrets in Postoperative Transsexuals; Canadian Journal of Psychiatry Vol. 34, February 1989, 43-45. Blanchard R, Steiner, B.W., Clemensen L., Dickey R. Gender Identity Clinic, Clarke Institute of Psychiatry, Toronto, Ontario, Canada.

more information
visit the site: Abstracts of the study
http://www.ncbi.nlm.nih.gov/sites/entrez?db=pubmed&uid=2924248&cmd=showdetailview&indexed=google

[2] CORRESPONDENCE BETWEEN 'ADAM' AND DR. RUSSELL READ

The correspondence exchanged between 'Adam' and Dr. Russell Reid. Adam's letter to Dr. Russell Reid clearly states explicit regrets.

more information
Adam's letter to Dr. Russell Reid, dated February 27, 1991
http://www.transgression.com/downloads/letter%20identity%20hidden,%20dated%2019900127.pdf
Reid's letter to Adam, dated March 4, 1991
http://www.transgression.com/downloads/letter%20Reid,%20dated%2019910304.pdf

[3] FITNESS TO PRACTICE PANEL HEARING IN THE MATTER OF DR. RUSSELL REID

Fitness to Practice Panel Hearing in the matter of Dr. Russell Warwick Stedman Reid.
more information
visit the site: letter from the GMC to Reid, dated May 25, 2007
http://image.guardian.co.uk/sys-files/Guardian/documents/2007/05/25/reid.pdf

[4] BAILEY'S TEST FOR AUTOGYNEPHILIA

In Michael Bailey's book, The Man Who Would Be Queen, @2003, Joseph Henry Press, ISBN 0-309-08418-0, Bailey provides a test to distinguish an autogynephilic transsexual from a homosexual transsexual. The test appears on page 193 of the book. Bailey's test includes a number of questions. The question are grouped, some have a value of +1; others have a value of -1. If the sum of the score gets to +3, Bailey instructs you to stop. The transsexual is autogynephilic. If the sum gets to -3, the subject is homosexual.

+1 As a child, did people think you were about as masculine as other boys?

+1 Are you nearly as attracted to women as to men? Or more attracted to women? Or equally uninterested in both? (Add 1 if "Yes" to any of these.)

+1 Were you over the age of 40 when you began to live full time as a woman?

+1 Have you ever been in the military or worked as a policeman or truck driver, or been a computer programmer, businessman, lawyer, scientist, engineer, or physician?

-1 Is your ideal partner a straight man?

-1 As a child, did people think you were an unusually feminine boy?

-1 Were you under the age of 25 when you began to live full time as a woman?

-1 Have you worked as a hairstylist, beautician, female impersonator, lingerie model, or prostitute?

more information
Bailey's faculty profile on North Western University website
http://www.psych.northwestern.edu/psych/people/faculty/bailey/

[5] GRÜSSER THALEMANN DIAGNOSTIC CHARACTERISTICS MODEL FOR IDENTIFY BEHAVIORAL ADDICTIONS IN PHENOMENON

Grüsser SM, Thalemann CN. Verhaltenssucht- Diagnostik, Therapie, Forschung. Bern: Huber; 2006.
more information
visit the site: Diagnostic Instruments for behavioral addiction: an overview
http://www.egms.de/pdf/journals/psm/2007-4/psm000043.pdf

[6] LONG-TERM FOLLOW-UP OF TRANSSEXUAL PERSONS UNDERGOING SEX REASSIGNMENT SURGERY: COHORT STUDY IN SWEDEN

Dhejne, C, Lichtenstein, P. Bern: Huber; Feb. 22, 2011.
more information
visit the site: Long-Term Follow-Up of Transsexual Persons Undergoing Sex Reassignment Surgery: Cohort Study in Sweden
http://www.ncbi.nlm.nih.gov/pmc/articles/PMC3043071/

Long-Term Follow-Up of Transsexual Persons Undergoing Sex Reassignment Surgery: Cohort Study in Sweden

CHAPTER 05

The Practice of Fear & Intimidation

It is 2:00 p.m. April 23, 2003. Psychology Professor Michael Bailey is presenting a lecture entitled "Gender Nonconformity and Sexual Orientation" to the Stanford University Psychology Department as part of its regularly scheduled departmental lecture series. Bailey's lecture follows shortly after the launch of his book, entitled, <u>The Man Who Would Be Queen, The Science of Gender Bending and Transsexualism</u>. As part of his presentation Bailey flashes a number of cartoons to warm up the audience. One of Bailey's cartoons is that of the effeminate gestures of a gay man in contrast to the macho gestures of a straight man. The audience laughs. However, as Bailey will soon learn not everyone in the audience is laughing.

Joan Roughgarden, a male-to-female transsexual and Biology Professor at Stanford University, covers the event for the Stanford Daily. Roughgarden transitioned at age 52 from being Jonathan Roughgarden to Joan Roughgarden. In fact Roughgarden reportedly went on sabbatical one day and returned to work as Joan. [1] Roughgarden's article is critical of Bailey for being insensitive to gays, lesbians and transsexuals. [2] From her knowledge of how the book is being received Roughgarden correctly reports that gay, lesbian and transsexual groups are organizing protests across North America in defiance of Bailey's book. In fact, Lynn Conway, a professor emeritus of electrical engineering and computer science at the University of Michigan at Ann Arbor, and Deirdre N. McCloskey, a distinguished professor of the liberal arts and sciences at the University of Illinois at Chicago were by this time aware of Bailey's book on transsexualism and had some discussion about the book's ramifications against their belief system surrounding the subject matter. In fact, prior to even reading Bailey's book, Conway was immediately suspicious of Bailey—who she knew to be a close associate of Ray Blanchard—and had planned to oppose any effort on Bailey's part to publish his ideas. In her e-mail message, dated April 10, 2003, Conway writes:

> I just got an alert about J. Michael Bailey's new book. It's just been published and of all places it's co-published by the National Academies Press, which gives it the apparent stamp of authority as "science" [....] As you may know, Bailey is the psychologist who promotes the "two-type" theory of

transsexualism [....] Anyways—not that there is much we can do about this—but we should probably read his book sometime and be prepared to shoot down as best we can his weird characterizations of us all.

Over the next two weeks, Conway and McCloskey read Bailey's book and begin to develop the argument that Bailey's book is inaccurate, insensitive and potentially damaging to transsexual women—a "vulnerable" community. The group's opposition to Bailey and his ideas are effectively preordained. In an e-mail addressed to GLAAD, dated April 21, 2003 Conway writes:

> This book will in time be viewed as very analogous to the Nazi propaganda films about Jews in WWII. It paints transsexual women as deviant, bizarre, pitiful figures and never shows the diverse reality of our true lives.

Bailey's book identifies a topology of transsexualism, delineated by sexual orientation. Bailey identifies otherwise heterosexual males as being autogynephilic and homosexual males as being homosexual transsexuals. Bailey relies on the concept of autogynephilia to define the types of transsexuals. Autogynephilia is the propensity of a male to become sexually aroused by the thought or fantasy of being or becoming a female. Bailey's book also goes into some length about nonconformity as being a psychological explanation of transsexual behavior. In fact, Bailey's book relies on the work of psychologist Ray Blanchard, who Bailey regards as brilliant, to support much of his arguments into the science behind transsexual behavior. Bailey's book includes the case studies of a number of transsexuals, who he had interviewed over the course of a ten year period dating back to 1994. Bailey does not state the claim that the book, itself, is scientific research, but rather Bailey relies on the interviews of male-to-female transsexuals coupled with scientific research to advance his ideas into the causes of transsexualism.

It is anyone's guess how Bailey's ideas insofar as a sexual motivation behind the transsexual behavior of one type of transsexuals incites hatred against transsexuals. In her e-mail, dated April 21, 2003, Conway fails to make the critical link. When in the seventies the homosexual movement successfully asserted itself against the greater population and professed the sexual interest towards members of the same sex is natural and not pathological, North American society largely adjusted in a non-violent manner towards the concept. If the transsexual movement were to distinguish itself similarly, if the transsexual movement advanced the role that sexual arousal plays in the early phases of transsexual behavior in lieu of the prevailing female narrative, the movement would gain more legitimacy (not less) in the eyes of the greater population, if for no other reason that the message is genuine and closely fits the reality of transsexual life stories.

Bailey's book does not explicitly incite hatred towards transsexuals or anyone for that matter. In all of Bailey's case studies he projects a degree of sympathy towards the people he interviews. The book will likely not go down in history as a greater threat to transsexuals than Hilter's book, Mein Kampf, was to European Jews. Yet, Bailey's book clearly falls outside of Conway and McCloskey's belief system. Conway and McCloskey's belief system—that a gender identity disorder and not a sexual motivation is the underlying cause of transsexual behavior and that the community of transsexuals is so diverse that it can't be explained by Blanchard's theories into autoeroticism and sexual paraphilias.

In 2003, Conway and McCloskey and other critics of Bailey levied claims that Bailey had committed ethics violations. In fact Conway and McCloskey directed up to four transsexuals to file complaints against Bailey to Bradley Moore, the vice president of research at Northwestern University. The letters appear to be authored by the same individual. Nevertheless, each claimed that they were used as research subjects without knowledge of the purpose of the interviews and without informed consent. In fact only two of the four transsexuals, who complained, appear in the book.

One of the four transsexuals, who complained, also made the allegation that Bailey had sex with her. The complainant, who is not identified, but is thinly disguised as Victoria a.k.a., Juanita, alleges that in 1996 Bailey gave the transsexual woman a letter supporting her "sex-reassignment surgery". She had the surgery in January 1997, was a guest speaker in one of Mr. Bailey's undergraduate classes the next month, and had sex with him at her apartment a year later. Bailey denies the allegation. Email records show that Bailey was at the home of his ex-wife with their children on the date of the alleged contact. Even so, it is unclear how an alleged sexual encounter between an author and an interview subject is a violation of ethics. Bailely's critics claim Bailey's relationship with Victoria was more than that of an author and an interview subject. Conway et al. claim that Bailey gave Victoria a reference letter to undergo a gender reassignment surgical (GRS) procedure, elevating Bailey to a position of authority. Bailey does give Victoria a referral letter to seek a GRS procedure. However, (as Bailey's critics correctly point out) Bailey is not a licensed psychologist in the state of Illinois. The letter from Bailey does not qualify as a letter from a licensed clinical psychologist that

Victoria needs to qualify for a GRS procedure, according to the standards of care. In Bailey's letter he does not evade this fact. There is no evidence that Bailey is (or ever was) Victoria's treating psychologist.

Northwestern University appointed Alice Dreger, Ph.D., an ethics scholar and patients' rights advocate, to investigate the allegations into Bailey. Dreger concluded that the allegations against Bailey were essentially 'groundless'. [3] Dreger found the allegations into misconduct could more accurately be described as 'harassment' and intimidation by Bailey's critics in an effort to destroy him personally and professionally. Andrea James, a transgendered advocate and close friend of Conway, downloaded images from Bailey's web site of his children, and published the images on her web site with sexually explicit content. In an e-mail James said that Bailey's work exploited vulnerable people, especially children and that her actions echoed that disrespect. [4]

In her findings Dreger writes:

> What happened to Bailey is important, because the harassment was so extraordinarily bad and because it could happen to any researcher in the field. [...] If we're going to have research at all, then we're going to have people saying unpopular things, and if this is what happens to them, then we've got problems not only for science but free expression itself.

On July 3, 2007 (or thereabouts) Kenneth J. Zucker, a researcher with the Centre for Addiction and Mental Health (CAMH), published Dreger's findings in the Archives of Sexual Behavior.

As soon as Dreger's report began to circulate, Conway, McCloskey and James reportedly attacked Dreger's findings as being 'biased' and "one-sided". In fact, James published a long hateful diatribe against Dreger, who she characterizes as a 'hermaphrodite monger'. On the site, tsroadmap.com, James publishes intent to launch further attack against Dreger in a passage that clearly includes ad hominem attack. James writes:

> [...] I will periodically begin the work of discrediting both Dreger's "partial history" and her. With luck, farsighted activists and academics will soon see - through Dreger's petty, self-aggrandizing nonsense and pathologization and see both for what they are. That's when Alice Dreger will finally get her wish. She will be a part of history, as an unfortunate historical footnote. [5]

The inquiry into ethics violations—which lasted almost a year—reportedly interfered with Bailey's research and adversely affected his reputation among his peers. In October 2004,

Dr. Bailey stepped down as chairman of the psychology department. He declined to say why, and a spokesman for Northwestern would say only that the change in status had nothing to do with the book. In a letter, dated November 4, 2003, Ray Blanchard, who worked closely with Bailey on the book, quit his position at the Harry Benjamin International Gender Dysphoria Association (HBIGDA), citing a decision by the HBIGDA to attempt to intervene in Northwestern University's investigation into the allegations against Prof. J. Michael Bailey.

Despite being fully exonerated by his peers, Conway continues to characterize Bailey as 'disgraced' on a page dedicated to the investigation into Bailey's alleged ethics violations. [6] Despite being fully exonerated by his peers, Conway elevates Bailey from a man whose career is in disgrace to a man disgraced. In effect, with each new tidbit of information Conway elevates Bailey to higher and higher levels of shame and ridicule. The page is akin to an *anorexic lounge of deflected thinking* where the head of the prey—in this case Bailey—is proudly on display on a mantle piece on the wall, following misreported fact after misreported fact. [7] Conway's strategy is thinly disguised—attack the ideas by attacking the man. Bailey is disgraced due in large part to Conway's interpretation and the interpretation of people that Conway invites into the discussion. Further, Conway publishes a complaint, dated May 10, 2004, she sent the University Provost. In the complaint Conway states she speaks on authority of the complainants. In fact Conway is not licensed to practice law and has no authority to speak for anyone other than 'herself'. On the contrary Conway's complaint against Bailey legally qualifies as contractual interference against his career and his means of earning an income.

Despite being fully exonerated, Conway, McCloskey and James have effectively silenced Bailey and who they regard as his 'cohorts' to the choir of the gender faithful, but not necessarily the world that matters—the greater population. A number of transsexual hailed Conway's efforts. Jamison Green writes:

> I think the Conway-led response had a positive effect on the community at large. I believe people felt empowered by it, because it modeled a powerful self-regard and courage to stand up for what one believes in, which is something that trans people need to see and internalize.

However, had Conway, McCloskey and James not acted, the institutions and colleges that regulate and govern scientific advancement would not have moved against Bailey's book. Bailey's book is an advancement of ideas—not necessarily science—and does not qualify for scientific scrutiny. Had Conway, McCloskey and James not acted, the book may have received some notoriety—the Lambda Society had considered the book for an award prior to the controversy—but not the notoriety it would eventually receive. Unfortunately, the book will forever be known as a flashpoint of controversy and an example of what happens when a person in the industry sees transsexuals outside a prescribed belief system. The over-the-top response by Conway et al. instilled a chill in the sexology field towards anyone daring to enter. Steven Pinker of Harvard University writes:

> The intimidation directed at Bailey will ensure that graduate students, post-docs, and other young researchers will not touch this topic with a ten-foot pole, starving the field of new talent. Only tenured professors who have decided to change fields—a tiny number—would take it on.

Given the relentless attacks against Bailey, Conway, McCloskey and James inadvertently alienated the greater population towards transsexuals. On August 21, 2007 Benedict Carey of the New York Times published a damning article into the behavior of Conway et al. [8] Conway elected not to be interviewed for the article.

Ironically, despite Conway's pleas towards the community to ignore the book, the controversy surrounding the book stimulated sales. The publisher reports that it sold 4,200 copies of the book, which is more than what the publisher expected from a trade book.

On August 22, 2007, after Dreger published her report, San Francisco's public radio station KQED hosted a program dedicated to the controversy surrounding Bailey's book. The broadcast follows an article published by the New York Times that reports on Bailey and the fact that he was exonerated by his peers. Michael Krasny of KQED engaged Bailey, Dreger, Roughgarden, and Mara Keisling the executive director of the National Centre for Transgender Equality. During the broadcast Roughgarden articulates that her continued beef with Bailey centers around a claim that the publisher advertised the book, itself, as scientific research and that it constitutes fraud. Bailey denies on air that the book, itself, is scientific research. Rather, Bailey maintains that the book draws from a number of interviews of male-to-female transsexuals coupled with scientific research led by Ray Blanchard of the CAMH. In fact the advertisement that Roughgarden quotes on air to illustrate her point does not say what Roughgarden interprets it to say. [9] The advertisement, which Roughgarden quotes on air, says it is about the scientific research and based on the scientific research, not that it is scientific research. Sadly, Roughgarden can't let it go.

Roughgarden's reports that her other beef with the book stems from allegedly concocting a narrative of a subject. The narrative is that of an effeminate gay man, who Bailey calls Danny. When Krasny challenged Roughgarden on the authenticity of the narrative by Danny, Roughgarden said she is not sure of whether it is true or not, only that it is reportedly not true. Roughgarden relies on a report from Anjelica Kieltyka, who claims that Prof. Bailey told her he fabricated the very end of the Danny story—a claim that Bailey denies. However, Roughgarden exaggerates Kieltyka's words to claim the entire narrative is a fabrication.

Lastly, Roughgarden reads passages out of Bailey's book, that deal with Bailey's observations of Latino She-Male sex workers, that she alleges is racist. In fact, Bailey correctly reports his observations that pre-operative homosexual transsexuals do gravitate to sex work. Bailer's observations—albeit limited—are indisputable. Bailey simply reports what he sees while 'cruising' a club known for homosexual transsexual sex workers. Whether the opinion that pre-operative homosexual male-to-female transsexuals gravitate to sex work is true or not is debatable. There is little research on the subject. In his work on the concept of partial autogynephilia Blanchard found that a significant number of heterosexual male-to-female transsexuals, who opt not to undergo a GRS procedure, have an erotic disposition towards sex work. Blanchard effectively found that money was not the primary motivator behind the interest these males had in sex work. Blanchard made similar observations about pre-operative homosexual male-to-female transsexuals. At the club Bailey witnesses a higher proportion of pre-operative male-to-female transsexuals, who gravitate to sex work, are Latino. If Bailey were to have visited the Web at 619 Yonge Street, in Toronto, Ontario in 1998 he may have come to the conclusion that a higher proportion of pre-operative transsexual sex workers, who gravitate to sex work, are Asian, due to the nature of Canadian Immigration policies.

What is particularly striking is each one of Roughgarden's beefs with Bailey runs counter to the evidence, yet Roughgarden, who is a Biology Professor at Stanford University, says them regardless. One would otherwise expect more discipline from a professor, who lectures in the sciences.

In a dramatic segment of the broadcast Ben Barres, a fellow professor at Stanford University, accuses Bailey and Blanchard of misrepresenting transsexuals and furthering oppression towards them. Barres correctly points out that transsexuals are a disenfranchised group of people and are broadly the victim of oppression. However, Barres fails to make the argument that Bailey's book advances the oppression of male-to-female transsexuals. Regardless, being the victim of systematic oppression does not give one (e.g., Conway, McCloskey, and James) a license to intimidate and harass people, who have a different opinion and advance different ideas no matter how personally offensive they may find their ideas.

Krasny reported that he had invited Lynn Conway to join the broadcast, but that Conway declined. Dreger reported that Conway asked for more time to formulate a response to her report.

Bailey is not the only academic to draw severe reaction, exercising academic freedom on the edge of political correctness. In 2006 Svend Hylleberg, the dean of the social science faculty of Aarhus University suspended Helmuth Nyborg, a Danish researcher, who reported a slight difference in the I.Q. levels between the sexes, on accusations of scientific malfeasance and fraud. [10]

Further, Conway, McCloskey and James are not the only transsexuals, who reportedly intimidate and harass researchers and academics. There are over one hundred loosely organized groups of transsexuals, who identify, intimidate and otherwise harass medical practitioners, researchers, academics, politicians and journalists, who operate outside of a prescribed belief system.

* * * * *

It is Thursday, March 29, 2007. Dr. Kenneth Zucker is speaking at an event sponsored by the Child and Adolescent Gender Identity Clinic at the CAMH. A number of medical practitioners and parents of children, who exhibit transsexual behavior, are in attendance. Zucker is an expert in gender dysphoria affecting children and adolescents.

A number of members of the transsexual health lobby of the Rainbow Health Network (RHN) are also in attendance. The RHN is a gay, lesbian, bisexual organization to promote health. Susan Gapka, a male-to-female transsexual, is the chairperson of the transsexual health lobby group of the RHN. [11]

The members of the transsexual health lobby are not in attendance on behalf of the interest of their children nor the health of their children. They have no children. The members of the transsexual health lobby are also not in attendance to add opinion to Zucker's speech. The event is not a round table discussion and none of the members of the transsexual health lobby is a medical expert. The members of the transsexual health lobby are there for one reason and one reason only—to instill fear into and intimidate medical practitioners, who work in sexology and other gender-related fields, to promote a political agenda.

The transsexual health lobby is suspicious of Zucker. Zucker has been working with children, who exhibit cross-gender variant behavior and the parents of these children to attempt to steer their children away from transsexualism, if possible. The transsexual health lobby is effectively attempting to interfere with the parents of children with affected gender dysphorias, the medical practitioners, who deal with such disorders, and the children, themselves. The transsexual health lobby believe they have a mandate to speak for children with affected gender dysphoria and if they do not, these children may not successfully complete their gender transitions and will in their opinion 'unnecessarily' suffer. The members of the transsexual health lobby clearly care about the medical treatment towards children, who exhibit transsexual behavior, but they are in no position to set down the standards of care. The transsexual health lobby is also suspicious of Zucker due to the fact that he gave academic credence to Dreger's report that exonerates Bailey.

The transsexual health lobby is also suspicious with the former Mike Harris government and the province of Ontario, who delisted gender reassignment surgical procedures from OHIP—a universal health care program. In fact they see Zucker in the camp of sexologists, who advocate that GRS procedures should remain outside the medical model, until more research is available. The transsexual health lobby wants the provincial government to reinstate GRS procedures as a publicly funded procedure immediately. The transsexual health lobby and its 'allies' are not interested whether the procedure is proven to be effective or whether the procedure is in the long term health interests for the people, who undergo it. In fact, the CAMH has never performed any qualitative study to determine how their former patients, who underwent a GRS procedure, are functioning in society five years post-operatively. The CAMH has no idea whether the people, who purported to feel more comfortable as members of the opposite sex and underwent the procedure, watched their lives get better or worse. The group is also not interested in the qualitative studies done in other jurisdictions, that found that a significant number of people, who underwent the procedure, no longer function as well adjusted members of the opposite sex. In fact, after learning a significant number of the male-to-female transsexuals, who were at one time ideal candidates for reassignment, had the same emotional problems post operatively, the gender identity clinic at Johns Hopkins University stopped offering the surgery. The transsexual health lobby, who are largely pre-operative male-to-female transsexuals, aren't interested in any more research towards transsexuality or the cause of transsexual behavior. The transsexual health lobby is only interested in immediate legislative action to provide universal access to GRS procedures and to make it available to all who want it.

After a short disruption of Zucker's speech, the members of the lobby and Zucker agree to take the lobby's 'issues' offline. The Rainbow Health Network scheduled a meeting with Zucker of the CAMH at Ryerson College, where they plan to voice their concerns and objections to him directly. The RHN scheduled the event at a later date at Ryerson College.

* * * * *

Self identifying transsexuals are rapidly becoming a political force. The era of transsexuals being at the mercy of the medical community to prescribe their medical treatment and chart the course of their gender transitions is over. The era of transsexuals being at the mercy of employers, who were otherwise quick to dispose of people, who were only different in the way they looked, dressed and acted, is over. The era of transsexuals being at the mercy of landlords, who evicted transsexuals from apartment buildings, is over.

Self identifying transsexuals have also served notice towards any would be journalists and story tellers of the transsexual experience—either 'correctly' portray us the way we envision ourselves or be the target of relentless intimidation and harassment.

Self identifying transsexuals are well organized. Transsexuals dialogue with other members of the community through online collaboration services, including list servers, blogs and chat services. Transsexual issue 'alerts' after reading or simply becoming aware of material that falls outside of a prescribed belief system or lends any credence to Blanchard's work. Over the course of debates and discussions—much of that online—self identifying transsexuals and the organizations that caters to the interests of transsexuals have developed a distinct political agenda. The political agenda involves changes in legislation that affect health, employment and access to services.

In the case of health, the prevailing agenda is to secure insured access to health services that aid in gender transition. The procedures of interest to male-to-female transsexuals differ from that of interest to female-to-male transsexuals. In the case of male-to-female transsexual the procedures of interest include, breast augmentation, gender reassignment surgical procedures, vaginoplasty procedures, facial reconstruction procedures and trachel shaves. In the case of female-to-male transsexuals the procedures of interest include mastectomy, and skin grafts to form a male penis.

The procedure of the greatest interest to male-to-female transsexuals is what is known as a gender reassignment surgical (GRS) procedure. The procedure involves a penectomy, that amputates the male penis, and vaginoplasty, that creates a neo vagina from the skin of the penis. The effectiveness of the procedure is questionable. For every study that shows the patients benefit from the procedure, there is a study that shows the procedure does more harm than good. None of the studies clearly shows that the procedure is a cure for so-called gender identity disorder. In fact there is no known cure for gender identity disorder. Males, who are diagnosed with a so-called gender identity disorder continue to suffer from the disorder despite undergoing procedure after procedure. Given there is no evidence that gender reassignment surgical procedure effectively ameliorates the symptoms of gender identity disorder, the procedure does not fall under the medical model to justify public funding for the procedure in jurisdictions that offer universal health care. Further, a gender reassignment surgical procedure is cosmetic in nature. The procedure does not touch the root causes of transsexual behavior, which are psychological. Given the procedure is cosmetic in nature and does not treat the psychological background, the procedure does not fall under the medical model to justify public funding for the procedure in jurisdictions that offer universal health care. The other procedures for male-to-female transsexual are similarly cosmetic in nature, and also do not fall under the medical model to justify public funding.

A gender reassignment surgical procedure is otherwise known as sex change surgery. Sex change surgery is a misnomer. There is no medical treatment or procedure that comprehensively changes one's anatomical sex. Despite the advances in medical science, no technique or procedure exists to change an individual's anatomical sex. Medical science has had some success altering a person's outward gender appearance through the introduction of synthetic hormones. Medical science has also had some success in cosmetically fabricating the anatomy of a penis into what looks like a vagina. Medical science has also had some success in cosmetically fabricating a male penis by way of what are little more than stents and skin grafts. These medical successes, however, are extremely limited. Further, the medical advances fall far short of satisfying the basic definitions of what is female of a species and what is male.

To illustrate the point in the case of a male-to-female transsexual, when he undergoes a penectomy and orchiectomy, he disables his ability to fertilize, which is a critical function of the male species, but he does not advance his ability to conceive or bear offspring, which are critical functions of the female of the species. In the case of a female-to-male transsexual, undergoing testosterone disables her ability to conceive, however, the attachment of a stent does not advance her ability to fertilize, which is a critical function of the male of the species.

In terms of employment, transsexuals work through GLBT rights organizations to introduce favorable changes to the human rights code, making it difficult for employers to discipline transsexuals or terminate their employment, if they fail to perform. People, who undergo gender transitions, spend enormous energies towards their physical appearance. People, who undergoing gender transitions, endemically have difficulties focusing on their work or become easily distracted. In fact a number of transsexuals have seen their positions terminated for performance related reasons. In many of these cases the transsexuals, themselves, recognize the reasons they were terminated were more to do with their performance at the job and did not submit grievances against their employers. Many employers terminate transsexual workers, who are in the midst of their transition, for performance reasons and performance reasons only.

* * * * *

In Canada, a number of transsexuals work through EGALE Canada to monitor and affect changes to legislation affecting the treatment of transsexuals. EGALE Canada places considerable spin on matters before the various Provincial Human Rights Commissions and legislation before the Provincial Assemblies and reports selective information to otherwise mislead the public.

When four pre-operative transsexuals launched a human rights action against, EGALE Canada covered the event. At the conclusion of the matter of Michelle Hogan, Martine Stonehouse, A.B. and Andy McDonald v. Her Majesty the Queen in Right of Ontario as represented by the Minister of Health and Long-Term Care EGALE Canada correctly reported that the Human Rights Tribunal of Ontario held that the government did discriminate on the basis of sex and disability against persons who had started medically-supervised transitions before October 1, 1998, and who received approvals for surgery from the Clinic within six years of having started their transitions. EGALE Canada also correctly reported that the tribunal held that the complainants should have received public funding for surgery, to allow them to complete the gender transitions they had begun at a time when public funding for surgery was still available. EGALE Canada also correctly reported that One dissenting member, Ross Hendriks, of the Tribunal would have held that the government's decision to remove public funding for sex reassignment surgery was discriminatory, arbitrary, reckless and an abuse of power. Hendriks wrote she would have ordered the government to fund sex reassignment surgery for all four complainants, since all four met the criteria for funding that had existed prior to October 1, 1998. However, EGALE Canada failed to give proper weight to the fact that a majority of the Tribunal held that the government's removal of public funding for sex reassignment surgery was not itself discriminatory. [12] [13] Instead in what is a poorly worded sentence, EGALE reports this fact but only in the context of admonishing the Tribunal for not providing reasons. The fact that the Tribunal held that the government's removal of public funding for sex reassignment surgery was not itself discriminatory is important and significant—more so than the opinion of one dissenting member of the panel. To illustrate the point open both documents (e.g., the one from the OHRC and the one from EGALE Canada) and compare how they report the decision and what weight they give to each fact. When you do, you will see how organizations like EGALE Canada applies spin to events.

The press release, dated November 11, 2005, references Laurie Arrons and Tammy Starlight as people to contact for further information. [14] Starlight, who resides in Vancouver, British Columbia, is a volunteer with EGALE Canada and sits on its board of Directors.

Based on the Tribunal's order—not withstanding the opinion of one member of the panel—the Ontario provincial assembly did not act to affect any legislative changes to reinstate public funding for GRS procedures. The commission has filed an appeal of the decision of the matter of Michelle Hogan, Martine Stonehouse, A.B. and Andy McDonald v. Her Majesty the Queen in Right of Ontario as represented by the Minister of Health and Long-Term Care to the Divisional Court.

EGALE Canada is one of hundreds of non-profit GLBT organizations that routinely publish selective and misleading information for purposes of advancing a political agenda.

* * * * *

Rosalyn Forrester is one person, who knows the human rights commission and how to profit on the unintentional mistakes of public servants. Forrester, who is (or was) the principle mind behind the now defunct, self-described advocacy site, Canadian Transsexuals Fight for Rights, has an extensive criminal record. On July 16, 1981, while living as Howard Spencer Forrester, the Peel Regional Criminal Court in Brampton, Ontario convicted Forrester of two counts of break and enter, and one count of mischief to private property and sentenced Forrester to twenty days of incarceration, two years of probation and a $250.00 fine. On November 9, 1983, the Peel Regional Criminal Court in Brampton, Ontario convicted Forrester of attempted theft and sentenced Forrester to three months of incarceration and a $750.00 fine. On October 1, 1985, the Metropolitan Toronto Criminal Court convicted Forrester of Driving Under the Influence (DUI). [15]

So when Forrester was improperly strip searched by a number of male officers of the Peel Regional Police, following arrests on May 29, 1999, August 11, 1999 and a third on March 8, 2001, over allegations of criminal harassment and violations of bail recognizance, Forrester did what Forrester arguably does best—he complained to the Ontario Human Rights Commission. [16]

The tribunal found the Peel Regional Police Services had indeed "humiliated" Forrester, but that the officers did not intentionally violate his rights. The tribunal ordered the Peel Regional Police to affect changes in their arrest procedures, respecting individuals, who are changing their physical anatomy to that of a member of the opposite sex, and produce a training video about transsexuality for the force. [17]

In fact Forrester and the Respondent reached a settlement insofar as remedies for 'mental anguish' prior to the final submissions before the tribunal and that the final submissions proceeded only to chart policy for the regional municipality that the Tribunal hoped would serve as a blueprint for other municipalities. In the decision the Tribunal instructed the Respondent to compensate the complainant an amount not to exceed $10,000.00 for 'mental anguish'. The Peel Regional Police had in fact drafted a policy of split searches on gender variant individuals prior to the submissions.

The final submissions went into some depth as to Forrester's psychological background. Given the nature of Forrester's complaint, the Respondent sought and was granted full access to Forrester's medical records of the psycho-therapy sessions with Dr. Toplack, who Forrester saw. Forrester reported feeling a sense of gender confusion at a young age and being uncomfortable seeing images of himself as male. Forrester also reported that he 'came out' in 1998. Forrester also reported that he felt attracted to women, that lead to some confusion with respect to his gender identity—in particular why he felt he was a woman, yet attracted to women.

Forrester testified that in April, 1999 he legally changed his name from his birth name, Howard Spenser Forrester, to Rosalyn Leslie Forrester. Forrester also reported being disabled and that he collected benefits from the Ontario Disability Support Program (ODSP).

Forrester reported that his troubles with the law started after ongoing difficulties with his common-law spouse. The police, responding to numerous complaints by Forrester's spouse, questioned, arrested and repeatedly strip-searched Forrester.

Forrester reported that during the arrest on May 29, 1999, that in order to accommodate his request for a female officer to strip search him, the Respondent provided him with a 'split search'. A split search involves an officer of one sex to strip search the suspect from the waist down and an officer of the opposite sex to strip search the suspect from the waist up. Forrester confirmed evidence by the Respondent that Forrester had advised the officers that he was "half way through a sex change" and also confirmed evidence by the Respondent that Forrester was wearing a sweatshirt with a bra underneath it. Forrester cooperated but felt humiliated by being seen by male officers. He reportedly cried.

A number of the officers that handled Forrester's arrest, including Constable Kevin Willson, did not know Forrester was undergoing a gender transition by his outward appearance and only became aware of that fact when Forrester told them of his gender transition. In fact, from the evidence Forrester was just beginning to undergo a gender transition at the time of the arrest and did not demonstrate any significant benefit from hormone replacement therapy. Forrester's driver's license also read Harold Spenser Forrester and not Rosalyn Leslie Forrester at the time of arrest.

Given the Respondent had reached a settlement with Forrester prior to the testimony, the tribunal did not pursue the evidence put forward by the Respondent that Forrester barricaded himself inside the store in which he worked at the time of the May 29, 1999 arrest.

Forrester testified that the police searched him a second time on May 29, 1999, at the (then) Clarence Street Court (Provincial Division), Brampton. The Clarence Street Court (Provincial Division), was replaced in 2000 with the A. Grenville and William Davis Courthouse. Forrester reported that this search was conducted by all male officers even though he asked for a female officer to perform it and said that he had had a female officer search him earlier. Forrester reported that four male officers searched him, and that they had to take his top off because he refused to remove his clothes. Forrester reported that the four officers stood back, stared and snickered. Forrester reported that one of the officers said, "So what are you?" Forrester reported that one of the officers told him, "Well, you'd better get used to it, because it's going to happen again today." Forrester felt despondent and his voice broke when he testified that, "I felt like ending it." None of the officers in attendance in May 29, 1999 corroborated Forrester's allegations. Again, given the Respondent had reached a settlement with Forrester prior to the testimony, the Tribunal did not attempt to assess the credibility of the various parties.

After a short bail hearing, the police released Forrester to the custody of his medical doctor, Heather Claire Davies, who acted as his surety. Davies is himself a male-to-female transsexual, who prior to his gender transition, practiced as John Hayden Morris. [18] When asked how he felt about the incident, Forrester reported that he felt he had been "raped, sexually assaulted." Forrester added that he felt, "like a freak in a zoo so they could stand there and go, ooh, ooh."

On August 12, 1999 the police conducted the third relevant strip-search prior to a scheduled remand date at court in order to change surety. Forrester reported that six male officers conducted the strip-search. Forrester testified that the six male officers made demeaning comments to him, such as, "oh, you liked it last time, you didn't mind last time," and "you still have a dick so we can still search you." Forrester reported that the male officers took his clothes off him in order to search him, and that the officers slammed him against a wall. Forrester reported that when the search was finished, the officers put him in a cell, and that the officers did not offer him any lunch. Forrester further reported that he broke down in his cell, feeling "completely done in." None of the officers in attendance on August 12, 1999 corroborated Forrester's allegations. Again, the Tribunal did not attempt to assess the credibility of the various parties.

There were also some inconsistencies in Forrester's testimony concerning the number of officers who performed the strip-search, varying in number from six in his testimony in chief, to four to five during cross-examination, and two to three in Forrester's Amended Complaint, dated October 6, 2003. The Tribunal also observed some unexplained silences or inconsistencies in his description of the manner in which the officers removed his clothing during the strip searches.

On March 8, 2001 at 5:00 a.m. the police arrested Forrester again for breach of the court order not to molest, harass or annoy his former spouse or child. Forrester reported that he was taken to 11 Division and held for a bail hearing. At that time, Forrester reported that male officers searched him, despite his request for a female officer. He said that four or five officers slammed him against the wall and tore off his top, bruising him in the process. Forrester's mother reportedly posted bail on Forrester's behalf. None of the officers in attendance on March 8, 2001 corroborated Forrester's allegations.

A Justice of the Peace released Forrester, on conditions, including a prohibition on contact with his former spouse. Forrester identified a motion record filed by his former spouse seeking a variation of the custody and access provisions of a previous court order, based on these allegations. Forrester reported that his (then) fiancée picked him up at the courthouse, and that out of fear, she immediately went into hiding.

Forrester reported that he "turned" himself in accompanied by his lawyer on March 18, 2001, at which time the police conducted a split search on him. In fact his lawyer arranged the strip search. Forrester alleged that the officers hurled more derogatory comments at him, including "hey, that's a guy," and that the male officers said to the female officer conducting part of the search, "I guess you drew the short straw." Forrester reported that the charges against him were withdrawn and that the Crown counsel was apologetic. That is not entirely forthright with the facts.

Counsel for the Respondent checked their records, and confirmed Forrester's evidence that there were two charges resulting from the arrest on May 29, 1999. One was withdrawn, on the condition that Forrester enter into a s.8(10) peace bond under the Criminal Code, supra, done without any admission of liability or guilt, and the second charge against Forrester was also withdrawn, and he was found not guilty. Forrester's appearance in in court on August 12, 1999 only dealt with changing surety. The police did not file any additional charges. A s.8(10) is part of the criminal code and is entered on a person's criminal record. The U.S. Immigration services refuses entry to people with criminal records including s.8(10) resolutions from entering the U.S.A.

Further, during cross-examination, it became quite apparent that the reason why Davies, who acted as Forrester's surety, had asked to be relieved was not due to retirement, as originally described by Forrester, but because Davies had lost confidence in him. Forrester's admission of the betrayal was reportedly painful.

Forrester brought some of the misery of the sequence of arrest after arrest onto himself by attempting to evade the police. The evidence suggests that the police unsuccessfully attempted to communicate with Forrester to serve Forrester with a restraining order. The restraining order prohibited Forrester from molesting, harassing and /or otherwise annoying his estranged wife and children. When the attempts by the police failed to locate Forrester, the court issued a warrant for his arrest. The police arrested Forrester on one occasion for no other reason than to serve Forrester with a restraining order.

Forrester launched a second human rights complaint as the first was being deliberated. In the second complaint Forrester alleges that the Mike Harris government acted in prejudice against him, when on October 1, 1999, the government delisted gender reassignment surgical procedures from OHIP. Although Forrester may have entered the program prior to the deadline, Forrester could by no means qualify to receive one prior to October 1, 1999.

Forrester is not the only transsexual to profit through the institutionalized complaint systems on often selective and misleading evidence. A number of transsexuals across North America have learned how to confound, intimidate, harass, and otherwise interfere with business affairs of former employers, public servants, and journalists for financial and/or other material gain.

references

[1] JOAN ROUGHGARDEN

Joan Roughgarden and her career are featured in Wikipedia.

[2] ROUGHGARDEN'S ARTICLE, PSYCHOLOGY LECTURE LACKS SENSITIVITY TO SEXUAL ORIENTATION

On April 25, 2003 the Stanford Daily publishes Roughgarden's article, Psychology lecture lacks sensitivity to sexual orientation.
more information
visit the site: Psychology lecture lacks sensitivity to sexual orientation
http://daily.stanford.edu/article/2003/4/25/psychologyLectureLacksSensitivityToSexualOrientation

[3] DREGER REPORT INTO THE MERITS OF THE ALLEGATIONS OF ETHICS VIOLATIONS OF MICHAEL BAILEY

On August 14, 2007 Dr. Alice Dreger. James publishes a final report into the merits of the allegations of ethics violations against Michael Bailey.
more information
visit the site: The Controversy Surrounding the Book, The Man Who Would Be Queen
http://www.bioethics.northwestern.edu/faculty/work/dreger/controversy_tmwwbq.pdf

[4] NEW YORK TIMES ARTICLE, CRITICISM OF A GENDER THEORY, AND A SCIENTIST UNDER SIEGE

On August 21, 2007 the New York Times published the article, Criticism of a Gender Theory, and a Scientist Under Siege, featuring Michael Bailey.

more information
visit the site: New York Times article, Criticism of a Gender Theory, and a Scientist Under Siege
http://www.nytimes.com/2007/08/21/health/psychology/21gender.html

[5] JAMES' CRTICISM OF DR. ALICE DREGER

On August 14, 2007 Andrea James publishes a hateful diatribe against Dr. Alice Dreger. James calls Dreger a 'hermaphrodite monger'.

more information
visit the site: TS Roadmap
http://www.tsroadmap.com/info/alice-dreger/hermaphrodite-monger.html

[6] CONWAY'S WEB PAGE OF THE CONTROVERSY SURROUNDING BAILEY'S BOOK, THE MAN WHO WOULD BE QUEEN

In 2003 Lynn Conway publishes a web page of links to information about the controversy surrounding Bailey's book, The Man Who Would Be Queen. Despite the fact that Bailey has been completely exonerated on ethics violations, Conway continues to paint Bailey as 'disgraced'. In fact largely the only people, who Conway reports to continue to hold Bailey in disrepute, are other homosexuals and transsexuals and their 'allies'.

more information
visit the site: Conway's page on the J Michael Bailey Investigation
http://ai.eecs.umich.edu/people/conway/TS/LynnsReviewOfBaileysBook.html

[7] THE ANOREXIC LOUNGE OF DEFLECTED THINKING

See Appendix B

[8] NEW YORK TIMES ARTICLE, CRITICISM OF A GENDER THEORY, AND A SCIENTIST UNDER SIEGE

On August 21, 2007 the New York Times published the article, Criticism of a Gender Theory, and a Scientist Under Siege, featuring Michael Bailey.

more information
visit the site: New York Times article, Criticism of a Gender Theory, and a Scientist Under Siege
http://www.nytimes.com/2007/08/21/health/psychology/21gender.html

[9] KQED'S ARCHIVE OF THE FORUM, DATED AUGUST 22, 2007

On August 22, 2007, San Francisco's public radio station KQED hosted a program dedicated to the the controversy surrounding Bailey's book. Michael Krasny of KQED engages Michael Bailey, Alice Dreger, Joan Roughgarden, and Mara Keisling.

more information
visit the site: The broadcast of the Forum, dated August 22, 2007
http://kqed02.streamguys.us/anon.kqed/radio/forum/2007/08/2007-08-22b-forum.mp3

[10] HELMUTH NYBORG

Helmuth Nyborg and his career as a researcher is featured in Wikipedia.

[11] THE TRANSSEXUAL HEALTH LOBBY OF THE RAINBOW HEALTH NETWORK

The Transsexual Health Lobby is a committee of the Rainbow Health Network. It sole mandate is to secure public funding for gender reassignment surgical procedure as part of the Ontario Health Insurance Plan—a universal health care model.

more information
visit the site: Transsexual Health Lobby | Rainbow Health Network
http://www.rainbowhealthnetwork.ca/transhealth

[12] THE EGALE CANADA PRESS RELEASE, DATED NOVEMBER 11, 2005

EGALE Canada's press release, dated, fails to give proper weight to the fact that a majority of the Tribunal held that the government's removal of public funding for sex reassignment surgery was not itself discriminatory in the matter of Michelle Hogan, Martine Stonehouse, A.B. and Andy McDonald v. Her Majesty the Queen in Right of Ontario as represented by the Minister of Health and Long-Term Care.

more information
visit the site: press release, dated November 7, 2005
http://www.egale.ca/index.asp?lang=E&menu=34&item=1182

[13] THE HUMAN RIGHTS TRIBUNAL OF ONTARIO DECISION IN THE MATTER OF MICHELLE HOGAN, MARTINE STONEHOUSE, A.B. AND ANDY MCDONALD V. HER MAJESTY THE QUEEN IN RIGHT OF ONTARIO AS REPRESENTED BY THE MINISTER OF HEALTH AND LONG-TERM CARE

Unfortunately, the Human Rights Tribunal of Ontario Decision on the matter of Michelle Hogan, Martine Stonehouse, A.B. and Andy McDonald v. Her Majesty the Queen in Right of Ontario as represented by the Minister of Health and Long-Term Care is not available online. The decision appears in the OHRC Annual Report for 2006.

more information
visit the site: Michelle Hogan, Martine Stonehouse, A.B. and Andy McDonald v. Her Majesty the Queen in Right of Ontario as represented by the Minister of Health and Long-Term Care (Tribunal Decision)
http://www.ohrc.on.ca/en/resources/annualreports/ar0607?page=eng-Tables_.html#Heading1170

[14] THE EGALE CANADA PRESS RELEASE, DATED NOVEMBER 11, 2005

EGALE Canada's press release, dated, fails to give proper weight to the fact that a majority of the Tribunal held that the government's removal of public funding for sex reassignment surgery was not itself discriminatory in the matter of Michelle Hogan, Martine Stonehouse, A.B. and Andy McDonald v. Her Majesty the Queen in Right of Ontario as represented by the Minister of Health and Long-Term Care.

more information
visit the site: press release, dated November 7, 2005
http://www.egale.ca/index.asp?lang=E&menu=34&item=1182

[15] ROSALYN LESLIE FORRESTER, A.K.A., HAROLD SPENCER FORRESTER'S CRIMINAL RECORD

Rosalyn Leslie Forrester, a.k.a., Harold Spencer Forrester, in fact has an extensive criminal record on CPIC.
more information
CPIC criminal record of Rosalyn Leslie Forrester page 1
http://www.acceptablelosses.ca/downloads/cpic_forrester_01.gif
CPIC criminal record of Rosalyn Leslie Forrester page 2

http://www.acceptablelosses.ca/downloads/cpic_forrester_02.gif

[16] THE SELECTION OF GENDER-BASED PRONOUNS TO DESCRIBE TRANSSEXUALS

In the book I utilize male appropriate pronouns to describe what are known as male-to-female transsexuals and female appropriate pronouns to describe what are known as female-to-male transsexuals. The reason is to provide clarity with respect to the person's biological anatomy.

It is legally possible in North America for individuals to change their sex code. The requirements differ from jurisdiction to jurisdiction. It is generally politically correct to respect transsexuals by using pronouns that are consistent with the gender they identify themselves to be. However, a person's legal sex and anatomical sex are not necessarily the same.

Despite the advances in medical science, no technique or procedure exists to change an individual's anatomical sex. Medical science has had some success altering a person's outward gender appearance through the introduction of synthetic hormones. Medical science has also had some success in cosmetically fabricating the anatomy of a penis into what looks like a vagina. Medical science has also had some success in cosmetically fabricating a male penis by way of what are little more than stents and skin grafts. These medical successes, however, are extremely limited.

Further, the medical advances fall far short of satisfying the basic definitions of what is female of a species and what is male.

To illustrate the point in the case of a male-to-female transsexual, when he undergoes a penectomy and orchiectomy, he disables his ability to fertilize, which is a critical function of the male species, but he does not advance his ability to conceive or bear offspring, which are critical functions of the female of the species. In the case of a female-to-male transsexual, undergoing testosterone disables her ability to conceive, however, the attachment of a stent does not advance her ability to fertilize, which is a critical function of the male of the species.

> Even the most 'passable' transsexuals have moments when they fail to pass. {Maxine Peterson, formerly Leonard H. Clemensen, M.A., CAMH, 2002]

[17] THE HUMAN RIGHTS TRIBUNAL OF ONTARIO DECISION IN THE MATTER OF ROSALYN FORRESTER VERSUS THE REGIONAL MUNICIPALITY OF PEEL - POLICE SERVICES BOARD

Unfortunately, the Human Rights Tribunal of Ontario Decision on the matter of Rosalyn Forrester versus the Regional Municipality of Peel - Police Services Board is not available online. The decision appears in the OHRC Annual Report for 2006. However, acceptablelosses.ca has made a copy of the decision available.

Rosalyn Forrester, a.k.a., Howard Spenser Forrester, had briefly made the document available on the (now defunct) site, Canadian Transsexuals Fight For Rights.

more information
visit the site: Rosalyn Forrester vs. Regional Municipality of Peel Police Services Board (Tribunal Decision)
http://www.ohrc.on.ca/en/resources/annualreports/ar0607?page=eng-Tables_.html#Heading1170
The Full Text of the Decision (word document): click here
http://www.acceptablelosses.ca/ContentManagement/Research/2006HRTO13.doc

[18] HEATHER CLAIRE DAVIES ' PROFILE ON THE COLLEGE OF PHYSICIANS AND SURGEONS

The College of Physicians and Surgeons of Ontario identifies Heather Claire Davies as John Hayden Morris in his/her profile.

more information
visit the site: Heather Claire Davies, M.D.
http://www.cpso.on.ca/Doctor_Search/former.asp?Type=ADV&intCurrentPage=1&sNameRefNo=0044571&i
CPSO=58549

CHAPTER 06

Feminism and the Feminist Treatment of Transsexualism

Feminism has had a long history of mistrust and suspicion towards transsexual people. Despite the fact that society recognizes transsexual women as female and has for some time, no major feminist organization explicitly recognizes transsexual women as women. However, this is on the verge of changing. Writers, activists, students and academics are all beginning to question rationale for continuing to exclude transsexual women. The major feminist organizations in the United States and Canada are actively debating the issue in open and closed forums.

Feminism is a political movement to affect the equality of the sexes.

The evolution of feminism is popularly broken down by a number of successive waves, drawing on the influences of other political movements. The civil rights movement of the sixties, the homosexual movement of the seventies, the anti-racism movement of the eighties all had influences on the feminist movement.

The first wave of feminism arguably began in the nineteenth century and ended in the early portion of the twentieth century. The first wave was confined to North America and Western Europe. The movement championed civil, legal and political rights for women. During this period feminists argued for and won the right to hold property and the right to vote.

The people behind the movement were largely white, educated, middle-class women acting in the interests of white, middle-class women. However, feminism had ties to the abolitionist movement. Women of colour were vocal about feminist issues even back then.

The second wave of the feminist movement arguably began in the mid-twentieth century. The exact beginning varies by region. In Canada, the second wave began around the time of the Royal Commission on the Status of Women in the mid-1960s. It is unclear where the second wave of the feminist movement ends and where to the third wave begins. It is generally accepted that the second wave blended into the third.

A number of feminists joined the civil rights movement, the peace movement and the environment movement and in so doing extended the scope of the feminist movement. In addition to fighting for legal, civil and political equality, feminism fought for improved working conditions for women, pay equity, the rights of self determination, (i.e., 'choice' over abortion), sexual freedom, and the political representation of women within a number of geographic spheres (i.e., third world countries, South America, etc.). At the same time the movement fought against domestic violence.

The movement during this period is characterized by the concept of sisterhood insofar as all women shared a common sexual and gender-based oppression. The slogan that was prevalent at the time was 'sisterhood is powerful'.

The third wave of feminism is not marked by a particular time per se. Rather; the third wave is marked by a stance or evolution of thought. The third wave has ties with the anti-racist movement, the homosexual movement, the anti-colonial movement. Women of visible minority groups, lesbians, immigrant women had enormous influence on the direction of the third wave of the feminist movement.

In addition to domestic violence and violence against women, Women in the third wave also fought against sexist stereotypes (e.g., women are supposed to be submissive, women are supposed to be more concerned with appearance, etc.).

"Third wave feminists are the first generation to grow up with feminist mothers. As we have grown up we have taken things for granted; things that our mothers never had. What we feel entitled to is completely different and the relationships we expect to have with partners is completely different." Krista Scott-Dixon says. Krista is a Ph.D. graduate of the Women's Studies programme at York University. There she concentrated on feminist theorist, feminist epistemology, methodology. Krista co-founded the site, www.trans-health.com, with Justin Cascio and Elaina Hardy.

The third wave is characterized by a shift away from the homogenous concept of sisterhood and towards the recognition of the difference amongst women. Women began to recognize that they had to concern themselves with oppression from within their own ranks. Depending on how women are positioned in the social hierarchy they can be both oppressors as well as oppressed.

"The changes we have seen in the past thirty years affect how we conceptualize women. It used to be people would regard women as one group. For example, there was the concept of sisterhood. And we are all sisters. Simply by the fact we are women. And now the sense is we have to look at differences. You can't assume you have a buddy-buddy relationship simply premised on the fact you are women. You have to deal with other real issues of class and race." Scott-Dixon says.

*   *   *   *   *

The issue of transsexuality was barely known, barely on the radar screen until visible male-to-female transsexuals began to knock on the doors of women's space. Renee Richards is one example. Renee Richards, a male-to-female transsexual, entered the Women's tennis circuit at age 40. Until that time it was felt that those people who look male had the gender identity of a male, and those that looked like female had the identity of a female.

Janice Raymond, a professor at University of Massachusetts, was one of the first feminist scholars to write about transsexuality. Unfortunately, for transsexual women fighting to be recognized as women, Raymond's book, The Transsexual Menace (1979), set the tone, and a very bad one, for the way feminists viewed transsexual people.

In her book, Raymond accuses male-to-female transsexuals of "raping" women's bodies invading women's space. She argues that transsexuals did not exist prior to modern hormone therapy becoming available and therefore are a new phenomenon created by the patriarchal medical community. However, there is ample evidence to suggest that transsexuals have existed as long as civilization has existed. Further, she denies the existence of cross-gender realities and bases her opinions solely on the position that sex and gender identity are fixed by chromosomes and the configuration of genitalia at birth.

Speaking of transsexual women, Raymond writes:

> If chromosomal sex is taken to be fundamental basis for maleness and femaleness, the male who undergoes sex conversion surgery is not female. (p. 10) Transsexuals are not women. They are deviant males. (p. 183).

"Raymond condemned male-to-female transsexuals, stating that anyone who was born male couldn't have the remotest idea of what it was to be a female. She refers to male-to-female transsexuals as artificial women." Eleanor MacDonald says. MacDonald is a professor in the Political Studies programme at Queen's University in Kingston, Ontario.

"She also condemned female-to-male transsexuals, stating that they were women who were in effect bailing out on feminism and not achieving their full female humanity." Macdonald says.

Raymond's book offers no explanation of how transsexuals come to realize a discomfort with their birth gender.

"I don't think she had a very a good understanding of transsexual people. I don't think she understands the whole phenomenon of transsexuality in general. The medical reason for it? Why people do it? Because of this lack of understanding, I think she imputes a hostile motive." Patricia Elliot says. Elliot teaches Sociology at Wilfrid Laurier University.

Despite what transsexuals regard as venomous language on the part of Raymond, a number of self-identifying transsexual women regard her book as having an element of truth.

"Unfortunately, those who might benefit most from them, i.e. transsexuals, tend to get so out of joint at her scornfully presumptive fallacy that they fail to see them. Which is no big loss, truly, since her ideas are no longer unique. However, there's one concept for which she has found appealingly succinct expression. It is a fundamental dilemma in the transgender community, and the most grievous fault she finds. It boils down to integration vs. integrity: integration as blending into society; integrity as loyalty to the truth of one's self." Nancy Nangeroni says. Nangeroni, a female-to-male transsexual, is the host of Gender Talk.

"Fortunately, this is not an idea which has escaped the transgender community. Today, more and more transsexuals refuse to lie about their past. More and more cross-dressers are owning up to others about their fascinating pastime. Finally, a community of transgender is emerging from the closet, reaping the benefits of brother- and sisterhood, not to mention self-respect." Nangeroni says.

Years later, Raymond was exposed for intellectual dishonesty for quoting a letter written by one of the subjects of her book, Angela Keyes Douglas, out of context.

Raymond went on to continue to be a thorn in the side of transsexuals in North America. She contributed to a study in the early 1980's on the topic of federal aid for transsexual people, seeking rehabilitation and health services. The paper effectively served to eliminate federal and some states aid for indigent and imprisoned transsexuals. It had a further impact on private health insurance which followed the federal government's lead in disallowing services to transsexual patients for any treatment remotely related to being transsexual, including breast cancer or genital cancer, as that was deemed to be a consequence of treatment for transsexuality. [Kay Brown, 1998]

* * * * *

The tone that Raymond set in the psyche of feminism found its way into hiring practices at a number women's volunteer organizations, including sexual assault centres, transition houses, International Women's Day coalitions, services for women in conflict with the law and women's centres to name a few. These organizations are what feminists refer to as women's space.

Sexual assault centres, the front line in the defense against male violence directed at women, are perhaps the most sacred of women's space. The staff are typically women—from the management staff to volunteer counselors. The rationale behind the hiring of women (particularly counselors) is simple. It is generally felt that women, who have undergone a violent act, prefer the counsel of a female in lieu of a male. However, what of transsexual women who are fully transitioned and legally female?

The Vancouver Rape Relief Society rejected Kimberly Nixon as a volunteer, peer counselor, not due to her gender. She is legally female. Rather, they refused to hire her due to the fact that she was raised as a male and in their opinion could not properly counsel victims of rape because she doesn't have the life history of a natal woman.

Ms. Nixon, distraught from her dismissal, hired Barbara Findlay, a lawyer and gay activist, and filed a complaint with the B.C. Human Rights Commission. In January, 2002, more than six years after the incident, the B.C. Human Rights Tribunal ordered the Vancouver rape crisis centre to pay Nixon $7,500.00 for injury to her dignity. The amount of the award is the highest in British Columbia for such a violation.

The B.C. Human Rights Tribunal also ordered the Vancouver Rape Relief and Women's Shelter to never again contravene the human-rights code with such conduct. Heather MacNaughton, who heard the case for the tribunal, ordered the society to cease denying jobs "for discriminatory reasons" to transsexuals.

Judy Rebick, the former head of the National Action Committee on the Status of Women, and one of Canada's leading feminists, defended the relief centre's action.

"The issue at stake is whether a women's group has the right to decide who its members are," she said in an interview after testifying at a hearing of a British Columbia Human Rights Tribunal.

"The challenge is 'who is a woman?'--which we're just beginning to deal with."

"What makes this tense—there's no question that transgendered people suffer from discrimination, they suffer a great deal. So, of course, [in] your heart as a feminist you want to be on their side in every fight but you can't because there is a conflict of rights."

Although the monetary amount was negligible, the decision had enormous implications to feminist organizations across North American. The message was clear. Begin discussions towards the inclusion of transsexual women or continue to face severe consequences in the courts.

In August, 2003 the Vancouver Rape Relief Centre brought the matter to judicial review—a form of appeal—and effectively overturned the ruling by the Tribunal. On December 19, 2003, the Supreme Court set aside the decision of the Tribunal, finding the Tribunal order was flawed. The Court ruled that the Vancouver Rape Relief Society had not discriminated against Kimberly Nixon and the group indeed has the right to freedom of association to organize as women only. The court further declined to send the matter back to the Tribunal for a rehearing.

Nixon took the matter to the British Columbia Court of Appeal. However, there she was again rebuffed. On December 7, 2005, the B.C. Court of Appeal held in a 3-0 decision that Vancouver Rape Relief has the right to prefer to train women who have never been treated as anything but female. The chief justice wrote in the endorsement: "The respondent Society was entitled to give preference to women who are not post-operative transsexuals, because there is a rational connection between the preference and the respondent's work or purpose."

On February 1, 2007, the Supreme Court of Canada refused to hear male-to-female transsexual, Kimberly Nixon's appeal from the decision by the British Columbia Court of Appeal that otherwise ruled in favor of the Vancouver Rape Relief Society, bringing Nixon's twelve year odyssey of litigation against the rape crisis centre to an end.

"We believe it is important for raped and battered women to have the choice of a women-only peer group for support. Now the Supreme Court of Canada has strengthened their right as well as strengthening our right to provide that support," says Suzanne Jay speaking on behalf of Vancouver Rape Relief. "This decision is important because it can affect many different groups especially those fighting violence and racism."

However, not all feminists agree with the rape crisis centre's policy.

"I think it is very problematic for that rape relief centre to say we don't recognize her as a woman and qualified to serve as a counselor because she does not have [the so-called, life experience of a woman.] They are using narrow criteria about who gets to be a woman, and who gets to be a woman in their view is only people who are designated to have been born women. Certainly Kimberly Nixon is not in that category. However, that doesn't mean she is not a woman." Patricia Elliot says.

Does the front line in the defense against male-aggression towards women require non transsexual female counselors? Can a male not suffice?

"I expect a sympathetic person to listen to me to understand what I have been through, not necessarily because they have been through it personally, [but rather] through the education program training program. To have some understanding of what a person is going through emotionally and socially and in other ways." Elliot says.

Barbara Findlay, Nixon's lawyer, who has overseen the case since Nixon's early petition to the BC Human Right Commission, sees the Court of Appeals decision as having much more gravity.

"Any group organized on the basis of disability, on the basis of religion, etc, and that provides services, can refuse to provide those services to anyone in the group that they don't like. They are free to discriminate." says Barbara Findlay. "Queers are particularly vulnerable to such discrimination. Disability groups have historically excluded people with HIV".

Is Nixon a champion of equal access? Is Nixon a champion of transsexual rights? Is Nixon a champion of the victims of rape? Nixon's true motivations are unclear. In any event it is impossible for Nixon to separate a genuine interest in the welfare of rape victims from her own needs for self actualization. The court ultimately upheld the rights for self-organized groups to self determine over the rights of the individual towards self-realization. To illustrate the point no male-to-female transsexual, including Nixon, can appreciate what it feels like to miss one's period after an incident of forcible rape, regardless of one's outward gender appearance.

The high court also ordered costs against for the application for leave of appeal. The prospects of the Vancouver Rape Relief Society successfully realizing costs will be difficult. Nixon has reportedly been on social assistance for nine years.

Findley reported that Nixon is currently working on historical restorations in the Vancouver area.

To Nixon's disappointment the courts will eventually have to rule on the matter of 'are they' or 'aren't they' another day and it will not involve "her."

* * * * *

Competitive sport is another space where natal women have historically guarded against male-to-female transsexual intrusions. The history of the exclusion of transsexual women from women's athletics began in 1968 when the International Olympic Committee introduced genetic testing to screen out, so-called, "female impersonators" from competing in women's events.

Although there were many rumors of transsexual women competing against women in women's events, including the story of two Soviet sisters, who won gold medals at the 1960 and 1964 Olympics, (then abruptly retired prior to genetic testing), there was in fact only one well-documented case. A German athlete named, Hermann Ratjen, assumed the name 'Dora' and competed in the woman's high jump in the 1936 Olympics. Ratjen, who taped his genitals to avoid detection from judges, confessed to his actions only years later. He claimed he was the victim of Nazi propaganda. He finished fourth.

Although the genetic test the IOC applied did screen out the intended targets (transsexual women and women with androgen insensitivity syndrome who are genetically male and have a slightly larger physique than women) it also screened out a number of biologic women, including women who gave birth to children (which intersexed people cannot do). In fact, the test screened out one in four hundred non-transsexual, non-AIS women. The IOC disregarded the test altogether for the 2000 Games in Sydney, Australia.

Regardless, the entry of transsexual women in women's sport remains an issue to this day. Michelle Dumaresq, a 32-year old, male-to-female transsexual, who earned a spot on the Canadian Women's Downhill Mountain Biking team, became the subject of great controversy. Dumaresq's teammates, Sylvie Allen, the reigning Canadian champion, and Cassandra Boon led a formal protest before the International Cycling Union, asking the ICU to ban Dumaresq from competing against women. Allen and Boon cited medical evidence they claim proves Dumaresq's male genetic predisposition gives her a physical advantage.

"I don't believe, personally, that it's fair," says Boon. "I just don't believe you can change the way that you were born and that you can completely turn yourself into the other sex."

The question of whether to regard the playing field as level seems to hinge on the degree of transition. In fact Dumaresq, who had reassignment surgery six years prior to earning a spot on the team, is a fully transitioned male-to-female transsexual. In fact medical experts are undecided as to whether a male with the hormone levels of a normal female pose any advantage over natal women. They cite there is simply not enough evidence to render an opinion one way or the other. In the case of Renee Richards, a judge ordered that Ms. Richards be allowed to compete with other females on the Women's Tennis Circuit on the basis of medical evidence, stating she had the hormone levels of a normal female.

Given there is no evidence to suggest that a genetic male with the hormone levels of a normal female pose any advantage over natal women, should Dumaresq not be given the benefit of the doubt?

"Yes." MacDonald says.

"When people transition—provided they take hormones—over time they lose the muscle mass they might acquire as a male. Other questions might come up about how long ago someone transitions and when the changes took place. We already know that among an average sample of men and women there are huge variations in hormone levels that have every reason to affect people performance."

*　*　*　*　*

Perhaps, no discuss of the systematic discrimination of transsexual women within feminist organizations would be complete without mention of the Michigan Womens Music Festival. The Michigan Women's Music Festival's policy on the participation of transsexual women is quite clear from its mandate. The festival's mandate reads as follows:

> "The Michigan Womens Music Festival is an event intended for women who were born and who have lived their entire life experience as female - and who current identify as women."

In the past the festival's marshals denied entry to transsexual women (who did not pass) and forced transsexual women (who successfully gained entry, but volunteered they were transsexual women) off the premises. Now the festival's marshals do not question a participant's sex status during the festival. However, if a participant volunteers that they are not female, they deny access to the individual. This "don't ask, don't tell" policy means that if no one volunteers their sexual history, anyone can participate in the woman-born women's event.

* * * * *

So why do a large number of feminists refuse to regard transsexual women as women? According to Scott-Dixon the answer may have something to do with the socialization process.

"Feminists, who oppose the inclusion of transsexual women, argue that transsexual women at one point in their life presented themselves as males. They feel that that experience shapes their perspective and identity forever." Scott-Dixon says.

The fear, that male-to-female transsexuals will bring male sensibility, or a sense of privilege, into women's space, may be relevant. However, the concern is hinged on what is known as a uniform life experience. Although a great deal is known about the socialization process of girls as they progress towards women, there is a great deal of variation. Class, race and ability all weigh into the development of a female child into an adult woman. Not all women are raised to be unassertive, dependent and nurturing. Researchers know even less about the socialization process of the transsexual child.

"If you are growing up and you are male embodied, you are going to internalize an awful lot of what is being told to girls. You may get a different set of messages. On the one hand [there is] a lot of pressure to grow up and become like a man. [...] on the other hand you are probably also internalizing some of the messages that are being giving to your sister or your female cousin or the girl next door about how you are supposed to behave. I don't know if we can say that children growing up, who are transsexual, are being raised as either male or female in terms of their social condition. There are probably more interesting combinations than that. It is hard to study because we don't know until well after the fact, but when you read people's autobiographies about what the experiences were like, it is pretty clear that they were not receiving the same information the same way." MacDonald says.

* * * * *

The issue of whether to include transsexual women in feminist models has produced two camps. The first camp regards male-to-female transsexuals as women in every sense of the being; the second continues to harp back to the philosophies of Janice Raymond. They do not regard male-to-female transsexuals as women but rather as somewhat troubled and perhaps envious individuals.

"If I had to divide people into two camps I would say that many of the people who are the second wave of feminists are more resistant to the inclusion of transsexual women to this community." says Scott-Dixon.

If male-to-female transsexuals are to be successful in their bid to win over the hearts of feminists, they will ultimately need to win over the legal system. To this end they are getting help from moderate elements of feminism.

Feminists had a huge influence on the advent of human rights both in the United States and in Canada. The articles, prohibiting discrimination on the basis of gender, was one of the first in the code. The code has protected women in the areas of equal access to housing, services and jobs. As the code matured it encompassed more disenfranchised groups. That is in keeping with the evolved feminist mandate—to fight against all forms of oppression.

For the sect of feminists who wish to continue to exclude male-to-female transsexuals, the argument is rapidly slipping from their fingers in the guise of human rights. In effect, they no longer own the argument of who constitutes being a 'woman'.

"The legal system is the lynch pin of the whole argument around gender. The legal system will force us to think about what we mean when we say the word, 'gender'. Obviously, discrimination on the basis of gender is prohibitive. But what do we really mean by that. I think these legal cases will force us to think what the discrimination on the basis of gender really means." Scott-Dixon says.

"What does it mean when someone, who looks like a man, says to me, 'I'm a women'. How do we deal with that? How do we understand that that is a real possibility and that can happen."

CHAPTER 07

Cross-dressing: Transsexualism in the Weak Form

Prior to 1995 Michael Gilbert was a little known tenured professor with the Philosophy Department at York University. Prior to 1995 Gilbert was known for augmentation theory and the theory of reaching agreement and little else. However, when one day in 1995 at the age of 50 he stepped out of his house wearing a blouse, a knee-length skirt and low pumps, and proceeded to lecture on gender and sexuality before one of his classes, Gilbert elevated his profile and thereafter could never go back to relative obscurity.

In 1995, Gilbert not only taught gender and sexuality, while cross-dressed, at York University, but also he spoke, while cross-dressed, at outreach workshops at York University. The outreach workshops—the Positive Space program—are designed to desensitize students to the reality that there are people who are sexually aroused to members of the same sex as well as the reality that there are men and women, who transgress gender norms. That same year Gilbert presented a paper, Cross-Dressing and Sexuality, to the International Congress on Gender in California. He didn't hide his identity. Gilbert admitted that he knew the risks.

Gilbert's public cross-dressing at University events began to attract attention. In 1998, Robin Wilson interviewed Gilbert over the telephone for an article she was writing for the Chronicle of Higher Education. In fact the publication sent over a photographer to take pictures of Michael cross-dressed that morning prior to the telephone interview. The publication asked only that Gilbert be cross-dressed for the picture. The Chronicle of Higher Education is a newspaper that is distributed to Universities and Colleges across North America. The subject of Wilson's article was transgender faculty and how transgender faculty have integrated their personal lives in the classroom. Wilson's article, <u>Transgendered Scholars Defy Convention, Seeking to Be Heard and Seen in Academe</u>, also features interviews with male-to-female transsexual, Deirdre McCloskey, female-to-male transsexual, Jacob (Jake) Hale, female-to-male transsexual, Ben Singer and male-to-female transsexual, Michelle Stanton. Gilbert is the only cross-dresser among the people interviewed. [1]

The Globe and Mail ultimately purchased the intellectual rights of the article and published the article, featuring Gilbert, on the front page of the February 14, 1998 edition of Focus. The editors changed the title of the article to read, To Sir--um Madam—With Love. At the time the article ran in the Globe and Mail, Gilbert was in Florida on holiday with his wife. In fact no one at the Globe and Mail gave Gilbert any warning they planned to publish the article that featured him. Gilbert learned of the article through his wife, who told him when she learned. Given Gilbert had worked for the Globe and Mail, contributing book reviews, Gilbert was disappointed, but later admitted that he would not have objected to the article had the newspaper warned him. On the contrary, Gilbert credits the article as shaping his notoriety. "On the positive side, the article made me the most famous transvestite in Ontario." Gilbert says.

Almost immediately after the Globe and Mail published the article, featuring Gilbert, a number of television producers contacted Gilbert to schedule appearances, including the producers of the Dini Petty show, Queer TV, Studio 2 and Rogers Cable.

Gilbert appeared on Studio 2 with Keith Green, who is also a public cross-dresser. Although the producers only planned to cover the topic for twenty minutes, Paula Todd, interviewed Gilbert and Green the entire hour. Gilbert reported that the audience was 'fascinated' with the topic. Gilbert reported the women in the audience were 'Okay' with cross-dressing provided their husbands didn't do it. Gilbert and Green not only promoted their cross-dressing life style on air, but also, they promoted Xpressions—a social club they founded with three other cross-dressers in 1995.

Gilbert was born in Brooklyn, New York, in 1945. Gilbert reported that his childhood was 'untroubled'. Gilbert was raised in a Jewish family. Gilbert's father was an accountant; his mother was a homemaker and part-time bookkeeper. Gilbert is the eldest of three children. He has a brother and a sister.

Gilbert earned an undergraduate degree from Lehman College in the Bronx. At the time Lehman College was known as Hunter College. Gilbert went on to take graduates studies at State University of New York in Buffalo, New York. However, prior to completing a degree from State University, Gilbert crossed the border into Canada to evade the Vietnam War. Gilbert reported that he was never in fact drafted, but rather, he wanted to avoid if at all possible any entanglements with the U.S. military actions in Indo-China. Gilbert eventually opted to become a Canadian citizen and abandoned his U.S. citizenship.

Gilbert, who is 62-years-old, has been married three times. Gilbert's marriage to his first wife lasted 14 years and ended in divorce. Gilbert's marriage to his second wife was brief. Gilbert's wife reportedly suffered from Hepatitis C. She died after the couple lived together for only two years. Gilbert credits a chat room for supporting him through his grief from the loss of his second wife. Gilbert's marriage to his third wife is reportedly going well. He optimistically refers to the marriage as 'never ending'. [2]

Gilbert is the father of one child from his first marriage. Gilbert's child is now an adult. Gilbert is also the step-father of four other children.

Gilbert reported that his proclivities towards cross-dressing began as a child. Gilbert reported that he masturbated, while wearing the clothing of the opposite sex, at a young age. Gilbert reported that, when he was going through puberty, he removed his female clothing after climax. Gilbert's practice of masturbation, while wearing the clothing of the opposite sex, continued into adulthood. Gilbert reported that as he grew older—around the age of 19 or 20—he no longer removed his female clothing after climax. Gilbert reported that when he cross-dresses, he often gravitates towards clothing that heightens his self awareness of being cross-dressed, including corsets, high heels, long nails, ruffles and lace. Gilbert, who is a relatively large man and wears a woman's size 16 shoe, reported that he has difficulty finding women's clothing that fits. Gilbert reportedly regards the word, panties, as the most erotic word in the English dictionary. [3]

In his article, In Praise of Fun, which appears in the Summer, 2001, edition of Transgender Tapestry, Gilbert reported that he gets 'sexually charged' through the act of cross-dressing. In his article, In Praise of Fun, Gilbert thinly disguises the activities that produce sexual arousal as 'fun'. Gilbert reported that he also enjoys reading what are known as short fetish stories, while cross-dressed. Gilbert finds fetish stories 'fun'.

Short fetish stories generally follow a common theme where a male, usually a young male, is forced to cross-dress and impersonate a female, for one reason or another, or is forced to physically become female. The short stories are fictional in nature and are often illustrated. Short fetish stories focus more on the being and not necessarily the clothing in isolation. In that regard the stories feed more into autogynephilic interests and not necessarily transvestic fetishistic interests.

Gilbert reported that cross-dressing for the purposes of being noticed is particularly enjoyable. Gilbert cites as an example a night when he and a number of other cross-dressers publicly cross-dressed at Casino Niagara and feared they were close to being kicked out for distracting the other patrons. It is not clear whether these outings provided the mental images that Gilbert relied on when he subsequently masturbated, while wearing the clothing of the opposite sex. Although Gilbert reports a great deal in the article, In Praise of Fun, Gilbert stops short of reporting his masturbation thoughts. However, Gilbert believes that his masturbation thoughts are not very different from the masturbation thoughts of other cross-dressers. Further, in his article, In Praise of Fun, Gilbert is quick to mention that the vast majority of the time he is cross-dressed he is not sexually aroused.

In fact, Gilbert's article in which he reports what he likes and doesn't like is not very different from what Dr. Ruth Westheimer reports in her sex columns. Gilbert's article, In Praise of Fun differs from Westheimer's articles only in that the subject matter is narrow. In his article, In Praise of Fun, Gilbert clearly demonstrates that he cares about the healthy sex lives of his readers and encourages his readers to enjoy cross-dressing for what it is and to not be ashamed of the practice. In reality cross-dressing for autoerotic purposes where the cross-dresser builds mental images from his public experiences to serve as masturbation thoughts in private harms no one.

Gilbert actively contributes a column for the Transgender Tapestry. Gilbert, who believes everyone has a responsibility to 'give back' to his/her community, chooses to write on transgender topics as his way of supporting the transgender community.

Gilbert is one of the few cross-dressers I interviewed, who was non-evasive about the role of sex in cross-dressing. When I challenged Gilbert about his openness, he responded:

> There is an expression in the gay community, the only thing that grows in the dark is a mushroom. I believe in openness. As a community we need to grow. We need to accept ourselves for who we are and how we came to being the way we are. If we accept ourselves as normal, we are normal. My whole shtick is to grow and learn and integrate with society.

When Toronto Life reporter, Macfarlane, reported on Gilbert in November, 2007, Macfarlane reported he saw nothing feminine about Gilbert. [4] Macfarlane writes:

> During our lunch, we engaged in a little baseball banter over the calamari (which proved to be excellent, just as he had confidently predicted). All in all, it would be hard to imagine a guy who is more of a guy. It wasn't until well into our meal, when I bent down to pick up my fallen napkin, that I happened to notice that his legs—quite shapely, actually—were shaved.

* * * * *

It is Saturday evening, October 25, 2003. A number of male cross-dressers and their significant others occupy a table at Zelda's. Zelda's is a restaurant in what is known in Toronto, Ontario as the gay ghetto. The restaurant is on Church Street south of Wellesley Street East. Zelda's is known for DRAG. Virtually all of the staff, including the waiters and bar tenders are cross-dressed.

'Maverick' and 'Leslie' are at the table. Maverick (not his real name) is an auto worker, who lives in Oshawa, Ontario. He is 35-years old (or thereabouts). Maverick is cross-dressed at the event. Leslie (not her real name) works as a clerk at a furniture store. She is a 32-year-old female (or thereabouts). Maverick and Leslie are legally married.

Leslie found Maverick via his web-site, Amanda Bower, in which Maverick publishes images of himself cross-dressed. In fact when she surfed onto Maverick's cross-dressing site, Leslie lived in San Antonio, Texas. After talking on the phone, Leslie flew to Oshawa, Ontario to meet Maverick. After a few months Leslie packed up her life in Texas and moved to Oshawa, Ontario to live with Maverick. A few months after that Maverick and Leslie wed.

Leslie is unlike many unsuspecting spouses of cross-dressers, who learn of their husband's proclivities towards cross-dressing after the relationship is consummated. Leslie deliberately sought a cross-dresser as a marriage partner. Leslie reported that as early as she can remember, she dressed her Ken's dolls in Barbie's clothing.

Maverick and Leslie now have a child. Maverick's proclivities towards cross-dressing are visible in the home. Maverick and Leslie reportedly allow their child to see Maverick cross-dressed.

Maverick and Leslie plan to publish images of the evening onto the page, adventures in cross-dressing, on Maverick's web site, Amanda Bower. In fact, Leslie and Maverick have been publishing images of their outings as 'Amanda' and Leslie for years. In fact, the couple shares their adventures with other cross-dressers, who connected with Maverick and Leslie through Maverick's cross-dressing site.

* * * * *

The Transsexual Delusion

It is Saturday, January 11, 2003, at 12:00 p.m. at a resort near Newmarket, Ontario off Davis Drive. The board of the cross-dresser social club, Xpressions, is meeting to discuss important matters. The club is privately operated. 'John' is among the group meeting at the resort. John (not his real name) is a public cross-dresser. John's femme name is Donna. Almost all cross-dressers have what are called femme names. Femme names are otherwise aliases that are appropriate (and not so appropriate) for members of the opposite sex.

John is an elected board member of Xpressions. At one time John acted as the club's president. In fact, John is renting a room at the resort over the entire weekend with 'James'. James (not his real name) is also a public cross-dresser and also an elected board member of Xpressions. James's femme name is Cynthia. Today, James is the acting president of the club.

John reportedly lives in a condominium complex in the Queen's Quay area of Toronto, Ontario. The condominium complex John lives in is high end and so is John. John owns and operates a successful distributorship that offers electrical motors. John's distributorship is located on Ellesmere Avenue in Scarborough, Ontario. Scarborough, Ontario is a suburb of Toronto, Ontario. Scarborough, Ontario is a short drive away from downtown Toronto, Ontario down the Don Valley highway. John also owns a boat, which he has moored at the Hamilton Royal Yacht Club. The Hamilton Royal Yacht Club is in Hamilton, Ontario. Hamilton, Ontario is fifty minutes from Toronto, Ontario down the QEW.

John, who is 63-years-old or thereabouts, reported that even at age 63 he still gets a "zing" from getting dressed as 'Donna'. John reportedly has been cross-dressing as Donna for most of his adult life.

John reportedly lives with 'Carol', a male-to-female transsexual. Carol (not her real name) reportedly is a member of the Canadian Coast Guard. In fact Carol is of Captain's rank. John and Carol have reportedly been an item for years. One could argue that John has everything a man could ask for, including a successful career, high end condominium, a live-in girlfriend (of sorts) and a boat on Lake Ontario.

John, however, downplays his sexual relationship with Carol. John reported the fact that he is involved with a male-to-female transsexual is coincidental. John reported that if things don't work with Carol, he vowed he will never get involved with a male-to-female transsexual again. In fact, all things being equal, the odds of randomly engaging in a relationship with a male-to-female transsexual in lieu of a natal female are 3,200 to 1. The odds assume that male-to-female transsexuals and natal women are indistinguishable. Anyone familiar with the

128

community of male-to-female transsexuals knows very few pass as natal females. In fact, even the most passable transsexuals have their moments when they fail to pass.

In 2005, (or thereabouts) John reportedly divested himself from his distributorship and retired from working. John reportedly had his boat physically moved to the Caymen Islands. John and Carol reportedly moved to the Caymen Islands to retire together on John's boat. In addition to the boat, John owns a timeshare in the Caymen Islands.

Unfortunately John and Carol broke off their relationship and John returned to Toronto, Ontario. On John's return to Toronto, Ontario the board of Xpressions re-elected John as president.

A former board member of Xpressions reported that John has a new girlfriend and that his girlfriend is, again, a transgender male. In fact, all things being equal, the odds of randomly engaging in two successive relationships with a male-to-female transsexual in lieu of a natal female are 10,240,000 to 1.

In October, 2007 John listed his timeshare in the Caymen Island for sale.

*   *   *   *   *

The current psychiatric opinion on cross-dressing is that the men, who periodically cross-dress, have what is known as an affected transvestic fetishistic disorder. A transvestic fetishistic disorder is the propensity for a male to become aroused by acting out in the clothing of the opposite sex. Public cross-dressing is a form of acting out. In each scenario, including Gilbert's evening at Casino Niagara, and the cross-dressers, who are seated around the dining table at Zelda's, none appears to be suffering from a disorder. In fact, Gilbert reported the evening at Casino Niagara as outrageous and enjoyable; the male cross-dressers and their significant others at Zelda's appear to be enjoying the evening. The men around the table at Zelda's who publically cross-dress likely do not see what they do as interfering with their lives to any great degree.

Psychologist Ray Blanchard of the Centre for Addiction and Mental Health (CAMH) has a different explanation towards the underlying motivation behind cross-dressing behavior. Blanchard found that all cross-dressers share a common background as children. He found that they all began to masturbate, while wearing the clothing of the opposite sex, at a young age—typically five years of age. In fact, male children begin to learn to masturbate typically at age five and in so doing begin to sexually develop. In fact, the majority of published life stories of cross-dressers and transsexuals report that they began to wear the clothing of the opposite sex at age five. Less often, the published life stories of cross-dressers and transsexuals report that they began to wear the clothing of the opposite sex at age eleven—the age when puberty begins. The correlation between the child's sexual development and emergence of gender variant behavior is near perfect and can't be dismissed.

Canadian male-to-female transsexual, Sylvia Durand, who lives in Ottawa, Ontario, reported 'feeling different' at age five. Durand is otherwise heterosexual and was married to a woman. Swiss male-to-female transsexual, Marie Noëlle reported her first gender-related drama occurred at age five. [6] Marie Noëlle reported she is heterosexual in that she had a sexual relationship with a woman. [7] Vancouver native, male-to-female transsexual, Michelle Dumaresq, reported her gender variant awareness began at age 'five'. Dumaresq reported she is female oriented, making her otherwise heterosexual. [8]

The age when males exhibit signs of gender nonconformity is radically different between autogynephilic males and males, who are reassigned at a young age. Janet Reimer reported that at seventeen months of age David Reimer, a.k.a., Brenda Reimer, ripped at the dress she attempted to fit him in. (see Chapter 3, the Decoupling of Sex and Gender) This distinction is extremely important to distinguishing a gender identity disorder from an autogynephilic disorder. A child's sexual identity is intact and immutable long before the child begins to sexually develop and experience sexual arousal.

Blanchard also found that all heterosexual cross-dressers exhibit a degree of autogynephilia. In other words, heterosexual cross-dressers derive sexual pleasure by acting out as women in public. Blanchard also found the thought of being a woman is a more powerful motivator to cross-dress than the thought of the clothing in isolation. The clothing of opposite sex comes into focus only through its contextual association. Although the regiment of wearing women's clothing is a small part of a female gender identity, it is fundamental.

Blanchard explains that as the autogynephilic child sexually develops he never fully externalizes the female love object. The autogynephilic boy effectively creates female space internally and begins to derive the benefit of sexual arousal by cross-dressing himself to appear female. In so doing the autogynephilic boy develops a love map between the thought of being female and sexual arousal.

If Blanchard is correct, all male-to-female transsexuals and cross-dressers share a common background and only differ in the degree of the behavioral addiction, underlying autogynephilia. Cross-dressing is transsexualism in the weak form. Cross-dressers isolate their daily lives from their sexual practices whereas transsexuals (both heterosexual and asexual) build their daily lives into their sexual practices.

Blanchard and the CAMH also studied the sexual interest towards transvestites, transsexuals and she-males. Blanchard refers to the phenomenon as gynandromorphophilia. Gynandromorphophilia is the propensity of a person—typically a male—to be aroused by a man, who changes into a female. In fact gynandromorphophilia is Greek for the love interest for the change from male to female. Blanchard found that gynandromorphophilia is a distinct sexual interest. Blanchard conducted his study of the subject in men who had sexual interest in transvestites, transsexuals and she-males. Blanchard's study did not look into the phenomenon in women.

Given 'Leslie' is a natal female, her interest in her husband's proclivities towards cross-dressing is rare, but not entirely unique. There are a number of women, who gravitate to cross-dressers and male-to-female transsexuals. In 1989, (or thereabouts) Patricia Aldridge, proprietor of the Take a Walk on the Wildside, married Veronica Brown, a male-to-female transsexual. In 1996, Aldridge married Tom Sloan, a.k.a., Roxy Wildside, a public cross-dresser. The Take a Walk on the Wildside is a self-described fetish boutique, catering to cross-dressers, transsexuals and DRAG queens. [2] The storefront at 161 Gerrard Street East offers a number of services for men, who wish to present as women in public, including makeovers, which Aldridge advertises as 'transformations'. The makeovers at the Wildside are expensive and run approximately twice what a woman would otherwise pay at Sears or the Bay at the Eaton Centre in downtown Toronto, Ontario. However, Aldridge has a somewhat captive market. Emerging public cross-dressers rarely opt for public exposure in the first few months of being out in public, if ever.

Despite being a public cross-dresser, himself, John's interest in male-to-female transsexuals is not uncommon. Blanchard found that 5% of all men prefer a male-to-female transsexual over a natal female and 31.1% of the men, who do, also cross-dress. Blanchard effectively identified a strong correlation between autogynephilia and gynandromorphophilia. [9]

A number of male-to-female transsexuals intimately partner with other male-to-female transsexuals. In fact the phenomenon is very common in otherwise heterosexual transsexuals.

Sylvia Durand and Cynthia Cousens are an example of a transsexual couple, each heterosexual, who are intimacy partners. Sylvia Durand, who is 43-years-old, is (or was) an active member of the Canadian Armed Forces. In fact Durand's career in the Canadian military spanned twenty years. Durand was born Sylváin Durand in 1966, or thereabouts. On January 20, 1998, Durand reported a psychiatrist diagnosed him as having a gender identity disorder. The next month Durand reported that he began to undergo hormone replacement therapy. Durand reported that he was married at the time and had a wife. Although Durand reported his wife supported him, he divorced his wife to be able to change his name and gender code. Durand, who reportedly knew he was different his whole life, reported he elected not to inform his wife about his proclivities towards cross-dressing until 1995 or thereabouts, when he felt it necessary. On July 1, 1998, Durand reported he began a real-life test. In August, 1999, Durand underwent a publicly funded GRS procedure from Dr. Yvonne Maynard at the Montreal Clinic. [10]

Cynthia Cousens, who is 56-years-old, is a retired police officer. Cousens also worked for the Victims Assistance Services of Ottawa—Carlton. Cousens was born, Peter Cousens, in 1951. Prior to his gender transition Cousens was married and had a wife. Cousens divorced his wife after 26 years of marriage to pursue a gender transition. On August 21, 2001 Cousens reportedly underwent a GRS procedure.

The transsexual couple lives in Ottawa, Ontario.

Child Psychologists are unclear whether exposing a child to autogynephilic cross-dressing is beneficial or harmful to sexual development. The conventional thinking in the area is that when the male cross-dresser acts out to his child, he is exposing what are otherwise sex toys to the child. The conventional thinking in the area is to not risk what is otherwise seen as potentially harmful to the child's sexual development.

* * * * *

You can read a lot into the sexual interest of a male cross-dresser by what he wears. If a cross-dresser routinely wears the clothing one would otherwise attribute to a street sex worker, chances are, the cross-dresser becomes aroused by the thought of being a female sex worker. If a cross-dresser routinely wears a traditional French maid outfit, chances are, the cross-dresser becomes aroused by the thought of being dominated or ordered about. If so, he has Masochistic Disorder. If a cross-dresser routinely wears the clothing one would otherwise attribute to a professional woman, chances are, the cross-dresser becomes aroused by the thought of being a female professional. York University Philosophy professor, Michael Gilbert, dresses as a professional female, when he lecturers on gender identity.

More and more cross-dressers are casually undergoing procedures to alter their gender appearance to enhance the experience of public cross-dressing. More and more cross-dressers are having laser electrolysis to protract their facial hair. More and more cross-dressers are casually undergoing hormone replacement therapy to develop breasts and alter the distribution of body fat to that of the opposite sex.

Cross-dressers can openly discuss their public experiences with other cross-dressers for hours. In fact a cross-dresser can discuss the mere act of walking down the street cross-dressed in public for hours and derive pleasure by simply reliving the warm sexual tension of the event.

* * * * *

The organizations that cater to cross-dressers are dramatically different than those that cater to transsexuals, which are largely political. Organizations that cater to cross-dressers are largely for social purposes.

Transgender support clubs, which used to reign supreme as the means of hob knobbing with public cross-dressers, are on the decline. Across North America transgender support organizations are seeing their numbers vanish. The membership of Xpressions, a Toronto-based transgender support club, is reportedly down 30% in 2003 from one year prior from 66 paying members to 46 paying members. Xpressions has less paying members in 2003 than it had at the end of its first year of operation in 1995. The club, Gender Mosaic, saw its membership steadily drop over the years. In 1996, the club had approximately 70 paying members. In recent years, the club dipped to a low of 25 paying members.

The club, Illusions, based out of Calgary, Alberta, also witnessed its membership figures drop over the years. The club has approximately 20 paying members in 2003. Eight years ago the club had over one hundred paying members. A small number of members broke off to open a Tri-Ess chapter, Serenity, however, it has out right folded.

The social club, 'Powder Puffs', based out of Orange County, California, out-right dissolved. Two years ago the club organized a cruise vacation on the Trans-Atlantic cruise ship, The Queen Mary. The event drew over one hundred people.

A number of chapters of the Society of the Second Self, (Tri-Ess) which used to have 1,100 members spread out to over 27 chapters in 1993, have also out right dissolved. Tri-Ess attempted to start chapters in Long Island, New York, and West Chester, New York, only to see the paying members absorbed into the New York Chapter and the chapters fold. The New York Chapter, buoyed by the collapse of Long Island and West Chester, has approximately 65 paying members. The membership level there has at best fluctuated. The South Jersey chapter, which is at the centre of a huge metropolitan area, has only a handful of members left and is barely in operation.

Social clubs are not alone in seeing their numbers dry up. Annual convention events, like Fantasia Fair, held in Provincetown, MA also saw its attendance decline markedly in 2002.

The value proposition of support organizations is several-fold. Firstly, support organizations offer access to a safe, social environment for the purposes of meeting like-minded people. Secondly, support organizations offer resources. Emerging cross-dressers thirst for knowledge about other people's experiences. Thirdly, the support organizations offer peer-level support, including spousal support. The married cross-dresser, whose marriage finds itself in jeopardy due to his proclivities towards cross-dressing, turns to support organizations, in hopes of saving his/her marriage.

The challenge the transgender support organization face is that members of support organizations typically do not find ongoing benefits from any one of these service offerings.

In the case of providing a safe, social environment, there are a number of reasons why cross-dressers no longer value this service on a perpetual basis. Firstly, the public is much more aware that there are men who cross-dress in public and men who change sex. Since 1993, and the advent of widely popular movies that featured transgendered characters, like the Crying Game, the social awareness of the existence of cross-dressers and transsexual has sky-rocketed. Not only has the media raised public awareness to the existence of transgender people, but also, the media has helped to soften how people regard transgenders and transsexuals. Today, almost everyone, who lives and works in a large urban area, knows someone who works with a transsexual. Today, almost everyone, who lives in a large urban area, has seen a cross-dresser in public. The shock value and stigma long associated with transgender is markedly less than it was ten years ago.

"The anxiety these people feel is the same as it was years ago." Tim Ayerst says. Ayerst is a professional therapist with a background in social work. "What has changed is the reception these people get when they venture out in the real world. They often find that there are no hassles being cross-dressed in public and that it is no big deal to be seen in public. Thus, once a cross-dresser develops a social network, he/she no longer needs the club to simply provide a safe, social environment to protect him/her from the real world." Ayerst says.

In the case of offering resources cross-dressers often consume a plethora of information within a year of being in public then their interest in resources wanes. "Cross-dressers are sick of attending lectures and watching people accept awards." Gord Murray says. Murray is a travel agent, who offers package vacations and other travel services to public cross-dressers.

"What cross-dressers want to do is dress and go out for an evening. [Understanding what cross-dressers want to do] isn't rocket science." Murray says.

In the case of offering peer-support, here too, the value of support suffers from the laws of diminishing returns.

"Transgender support groups are by their nature victims of diminishing returns." says 'Christine', a 54-year-old, transgender individual. Christine (not her real name) is also a member of Tri-Ess, the largest support group for heterosexual cross-dressers. "As a person or couple deal with the stress and confusion associated with being transgendered or married to a transgendered [individual] and become involved in a support group, they have access to voluminous information and a social setting. Through their involvement with a support group eventually one of two things occurs: 1) the acceptance level of the spouse grows and the relationship takes on a new dimension along with the transgender person finding a new understanding of themself; 2) the spouse rejects the concept and the relationship ends. In either case the 'need' that brought them to the support group in the first place evaporates."

Whereas transgender support organizations appear to be in decline, the loose organization of vacationing cross-dressers appears to be the wave of the future. The loose organization of vacationing cross-dressers organize word-of-mouth events that target other like-minded cross-dressers, include Diva Las Vegas [Las Vegas, Nevada], Pink Fest [Chicago, Illinois], and Eureka Enfemme Getaway [Eureka Springs, Arkansas] to name a few. These loose organizations tend not to be affiliated with any non-profit, transgender organization. Rather, these loose organizations tend to compete head-to-head with conventions, like Southern Comfort, IFGE and Fantasia Fair and do so successfully.

None of these events requires participants to make reservations, nor pay entrance fees. They are pretty much pay-as-you-go.

"What we are seeing is venues like Diva Las Vegas, and cruise tours sky-rocket in popularity." Murray says.

"If you attend Diva Las Vegas and want to go to see an Elvis impersonation act, you can go with a large group of cross-dressers. If you attend Pink Fest and want to go to a DRAG show, you can go with a large group of cross-dressers. You don't have to pay a registration fee to simply be there." Murray says.

There are also a number of travel services that have gotten into the act. Dignity Cruises specifically markets vacation packages to cross-dressers and their partners.

If transgender support organizations wish to fight attrition, they simply have to offer a value proposition that addresses ongoing needs. One need that resonates to individuals who consider or actively pursue transition is human rights as there is a huge opportunity to right social inequities directed towards transsexuals. The organization that provides political services (e.g., including lobbying services like GenderPAC) will survive so long as there are social inequities and people determined to fight them. In other words, transgender support organizations have to become more political, if they want to be anything other then a club for transient emerging cross-dressers. They must act as the voice of transgender individuals in major metropolitan areas and show a presence. However, presence runs counter to the public cross-dressing lifestyle, which is largely discrete.

references

[1] THE CHRONICLE OF HIGHER EDUCATION ARTICLE, TRANSGENDERED SCHOLARS DEFY CONVENTION SEEKING TO BE HEARD AND SEEN IN ACADEME

In 1998 The Chronicle of Higher Education published the article, Transgendered Scholars Defy Convention, Seeking to Be Heard and Seen in Academe. In Saturday, February, 1998, the Globe and Mail republished the article on the front page of the Focus section.
more information
visit the site: The Chronicle of Higher Education article, Criticism of a Gender Theory, and a Scientist Under Siege
http://chronicle.com/colloquy/98/transgender/background.htm

[2] THE TORONTO LIFE ARTICLE, THE NAUGHTY PROFESSOR

In November, 2007 David Macfarlane of The Toronto Life Magazine, publishes the article, The Naughty Professor. Macfarlane's article features Philosphy Professor, Michael Gilbert.
more information

visit the site: The Toronto Life Article, The Naughty Professor
http://www.torontolife.com/features/naughty-professor/

[3] THE TRANSGENDER TAPESTRY ARTICLE, IN PRAISE OF FUN

The Transgender Tapestry published Michael Gilbert's article, In Praise of Fun, in the Summer, 2001 edition of the magazine. In his article, In Praise of Fun, Gilbert discusses what he as a cross-dresser finds erotically pleasurable. In his article, Gilbert effectively equates 'fun' to erotically pleasing.

more information
visit the site: The Transgender Tapestry published Michael Gilbert's article, In Praise of Fun
http://www.ifge.org/magazines/94_summer01.htm

[4] THE TORONTO LIFE ARTICLE, THE NAUGHTY PROFESSOR

See [2]

[6] GENDER TALK BROADCAST #255

In his broadcast, dated April 24, 2000, Nancy Nangeroni interviewed Sgt. Sylvia Durand of the Canadian Armed Forces.

more information
visit the site: gender talk broadcast #255
http://www.gendertalk.com/radio/programs/251/gt255.shtml

[7] CONWAY PROFILE, MARIE NOELLE

Lynn Conway features Marie Noëlle as a transsexual success story.
more information
visit the site: Conway Profile, Marie Noëlle
http://ai.eecs.umich.edu/people/conway/TSsuccesses/Marie-Noelle.html

[8] SPORTS ILLUSTRATED ARTICLE, A LEVEL PLAYING FIELD

In his article, A Level Playing Field, in the June 24, 2003 edition of Sports Illustrated Mike Fisk reports on Michelle Dumaresq.
more information
visit the site: Sports Illustrated article, A Level Playing Field
http://sportsillustrated.cnn.com/inside_game/mike_fish/news/2003/06/24/fish_dumaresq/

[9] BLANCHARD STUDY ON GYNEANDROMORPHOPHILIA

Ray Blanchard and Peter Collins conducted a study to explore gynandromorphophilia, that is, the sexual interest in crossdressed or anatomically feminized men. Blanchard and Collins coined the term, gynandromorphophiles, to designate males with distinct interest in feminized males, including men wearing women's clothing and men that have altered the contours of their body to a more feminine appearance by either surgical or hormonal means, but have male genitalia intact. Prior to the study performed by Blanchard and Collins, Money and Lamacz noted the existence of such men in a working paper published in 1984.

Men with sexual interest in transvestites, transsexuals, and she-males; Journal of Nervous and Mental Disease, 181, 570-575 (1993), Blanchard R, Collins PI. Gender Identity Clinic, Clarke Institute of Psychiatry, Toronto, Ontario, Canada.
more information

visit the site: Abstracts of Relevant Articles by Blanchard
http://www.autogynephilia.org/Abstracts.htm

[10] GENDER TALK BROADCAST #255

See [6}

CHAPTER 08

The She-male Phenomenon

It is Tuesday evening, August 16, 2005 at Clinton's on Bloor Street West in downtown Toronto, Ontario. Allan Novak and The Joe Blow Show Production company is producing a pilot for the CTV comedy show, PUNCHED UP. This evening in addition to thirty other actors, Novak has hired 31-year-old Nina Arsenault, a.k.a., Rodney Arsenault, for a principal role in the production. Arsenault is otherwise a male-to-female transsexual, who either identifies or had identified at one time as a she-male.

Arsenault is well known in the television industry. A number of the cameramen speak very highly of Arsenault. One cameraman in particular applauds Arsenault as being a "real pro." He is, including his activities beyond acting. Tonight Allan Novak, the producer, asks Rodney to nonchalantly serve drinks to two other principal actors on stage as if his transsexual presence blends into the ambiance of production with little or no attention paid to him.

Arsenault has been working in television since he graduated with a second Masters degree in Playwriting at York University in 1999. Prior to that in 1998, Arsenault completed a Master's degree in Theater Directing at University of Capetown in Capetown, South Africa. While studying at the UCT, Arsenault directed a number of plays at the Drama Department, including "Lion in the Street" by Judith Thomson, and the "Big White Room", which Arsenault wrote. In 1996, during his undergrad program, Arsenault began to take steps to explore what he regarded as a gender dysphoria. At the time Arsenault had the intention of eventually undergoing a gender reassignment surgical (GRS) procedure and living the rest of his life as a male-to-female transsexual. Arsenault applied for an assessment at the Gender Identity Clinic at the Centre for Addiction and Mental Health (CAMH) (formerly known as the Clarke Institute of Psychiatry). Arsenault waited six months or thereabouts for the assessment. On the day of the assessment Arsenault was seen by five experts in total, including three psychiatrists and two clinical psychologists. The assessment lasted an entire day. Dr. Ray Blanchard, Ph.D. was one of the two psychologists to examine Arsenault.

Blanchard reportedly saw nothing feminine about Arsenault. In fact, Blanchard was reportedly less than impressed with Arsenault's transsexual proclivities and attempted to dissuade the young man in front of him from any gender transition. Blanchard reportedly told Arsenault that he could effectively live as an effeminate male with little gender discomfort and that if it were up to him he would not approve Arsenault for a GRS procedure. When Arsenault questioned Blanchard to the reason for his objection, Blanchard reportedly said that a gender transition is "a lot" to go through and the surgery, itself, is "messy."

When the group of five experts learned that Arsenault had interests in an acting career, they—one by one—were opposed to any gender transition. However, when Arsenault reportedly corrected them that he had plans to direct, they were more amenable to a future gender transition. The group of five experts were—one by one—apparently of the opinion that a transsexual could never gain any foothold in acting. Arsenault, himself, now admits that the opportunities for work in acting as a visible transsexual are not sufficient to base a career on.

Arsenault received the final report from the group at the CAMH several months after the assessment. To Arsenault's surprise the final report ran counter to Blanchard's negative evaluation. The group recommended that Arsenault complete his master's degree at York University, and if Arsenault was still interested in a gender transition, he undergo hormone replacement therapy with endocrinologist, Betty Chan, M.D.

Although Arsenault characterized his experience at the CAMH as 'incidental' in the context of his gender transition, Arsenault followed through on the report.

With his graduate studies complete, in 1998 Arsenault began undergoing hormone replacement therapy with endocrinologist, Dr. Betty Chan. Chan put Arsenault on tricycline 20 mg o.d.. Ortho-Tricycline is widely known as 'the pill'. Women otherwise take tricycline for birth control. Chan advertised the protocol as all Arsenault would require to complete a gender transition. Arsenault reported that in fact the dosage was so low that he failed to benefit from the medication.

Chan's examinations of Arsenault were reportedly 'bizarre'. Arsenault reported that on each visit Chan had Arsenault remove his shirt. With his shirt off Chan reportedly felt Arsenault's nipples for prolonged periods of time and proceeded to ask Arsenault questions. Arsenault reported that on each visit Chan also directed Arsenault remove his pants. With his pants removed Chan reportedly felt Arsenault's testicles. Arsenault reported that Chan asked

Arsenault if he was becoming 'passable' in public. Arsenault reported that he didn't feel the question was relevant.

In 1999, after one year with Chan, Arsenault reportedly switched to general practitioner, Dr. John Goodhew, on the advice of members of the community of male-to-female transsexuals. Goodhew questioned tricycline as a feminizing agent. Goodhew reportedly said words to the effect that 'that won't do anything'. During his first visit with Goodhew, Arsenault reportedly asked him if he wished him to take off his shirt to examine his breasts. Goodhew reportedly said 'no', questioning why Chan would want a male-to-female transsexual to do that. However, Goodhew also reportedly asked Arsenault if he was becoming 'passable' in public. Arsenault reported being intimidated by the question. [1]

"As a transsexual I felt this pressure to be passable, then I wanted to be on more hormones cause I thought that could make me more passable." Arsenault says.

In 2004 or thereabouts, Arsenault reportedly left Dr. Goodhew in favour of Dr. Keith Loukes of the Sherbourne Health Centre, who worked with Arsenault at PrideVision. When Loukes is not available, Arsenault reportedly sees Dr. Tam of the Sherbourne Health Centre.

Arsenault reported that he circumvented his doctors to obtain feminizing medications on at least one occasion. In fact, Arsenault reportedly obtained one prescription from a doctor who he otherwise had an intimate relationship with. Arsenault also reportedly received prescription medication from visitors of an online pay-per-view site, who were also licensed practitioners.

Arsenault reported losing interest in undergoing a GRS procedure when in 1998, he was contacted by Mitchel Raphael, a reporter at the National Post [4]. Raphael was following up on the story of the Mike Harris government cutting public funding towards GRS procedures and wanted to speak to male-to-female transsexuals, who wanted to live as females, but were not interested in having genital surgery. At first Arsenault declined the interview, stating that he didn't qualify for the interview. After a week, Arsenault re-evaluated the role of a GRS procedure in his transition and contacted the reporter, saying that in fact he did qualify for the interview.

"I hadn't examined it fully." Arsenault says.

Arsenault reported undergoing upwards of forty-seven procedures—many of them involving invasive surgery—towards his gender transition. Arsenault characterized the surgeries as otherwise removing his male features.

Arsenault reported that he began undergoing his gender transition while attending York University, which would place the first procedure in 1998. In 2001 Arsenault went to Dr. Douglas Ousterhout in San Francisco for what is known as facial feminization surgery. Under the care of Dr. O (as he is known by the community of male-to-female transsexuals), Arsenault underwent a chin reduction, forehead, thyroid cartilage reduction, brow re-position, and rhinoplasty procedure. The five procedures—which Arsenault underwent in one visit—set him back $45,000.00. Arsenault reported that he only now acknowledges that the cost of the five procedures was expensive.

Arsenault speaks highly of Dr. Ousterhout. Arsenault says that Dr. O. is very careful to hide or diffuse the scars of surgery wherever possible. Arsenault credts Dr. Ousterhout as having extensive experience treating males, who took steroids and sought to have the pronating effects of the steroids reversed.

In 2005 Arsenault underwent an orchiectomy in Guadalajara, Mexico from Dr. Antonio Ramirez Chavez, a.k.a., Dr. Sunny. Arsenault reported his primary motivation was to keep his testosterone level as low as possible. Prior to undergoing an orchiectomy, Arsenault was taking Androcur 300 mg o.d. and Estrace. Arsenault reported that androcur and estrogen poses high risks in terms of blood clotting. In one of his articles in Fab magazine Arsenault credits George Bush and a concern over economic collapse and a potential disruption in the supply of pharmaceutical products for his decision to undergo an orchiectomy.

"I like know that [...] my testosterone level won't come back no matter what happens, so if there is another nine eleven and there's this national crisis [...] you can't get industrial dosages of hormones, you [will] be Okay. " Arsenault says.

Arsenault reported that Mexican cosmetic surgeon, Dr. Antonio Ramirez Chavez, a.k.a., Sonny, is one of the few cosmetic surgeons, who performs rib surgery, where instead of removing the lower rib, the doctor pushes the rib inwards to be less apparent.

As Arsenault underwent surgical procedure after surgical procedure, his objectives slowly changed. In the beginning Arsenault reportedly underwent surgical procedure for the purposes of passing as a woman in public. For Arsenault looking like a woman or being confused to be a woman was the pinnacle. Later, Arsenault reportedly sought surgeons, who were amenable to go beyond standard North American results and who would accommodate a more cosmeticized result. In the case of his lips Arsenault reportedly asked his surgeon to give him lips that look feminine, but then to go a little farther and then go well beyond that. With each procedure Arsenault pushed himself farther and farther away from the "standard" female—if such a concept exists. In any event Arsenault reported that "standard" female concept does in fact exist at least in his mind. Further, with each procedure Arsenault reportedly created a more plasticized look. At the some point Arsenault reportedly went through a phase of identifying as a she-male and not necessarily a woman or transsexual. Arsenault reported, however, that his self-identifying she-male phase was brief.

In 1999 after graduating from a second master's program at York University, Arsenault took a position as an Acting instructor at York University. At age twenty-four Arsenault was the youngest University instructor in Canada. To say Arsenault was an over achiever is perhaps an understatement. Arsenault had by this time directed plays on three continents.

"It wasn't like I couldn't make it as a guy." Arsenault explains.

Arsenault reported that he just wanted to live as a woman. As Arsenault began to transition his life took a dramatic turn. Arsenault reported spending enormous mental energy and money on his transition. Arsenault reported spending over $100,000 .00 on surgical procedures. To afford the numerous surgical procedures, Arsenault turned to online stripping on a pay-per-view site operating at 619 Yonge Street. Arsenault reportedly promised himself that he would stop, after paying off all the surgical procedures he planned to undergo. That never happened. Arsenault reported that he continues to strip online for paying subscribers and that he continues to contemplate undergoing the knife.

Arsenault reportedly also engaged in escorting—a form of sex work—while teaching at York University. Arsenault took money from wealthy men in North America to travel to meet them. Arsenault insists that he never had sex with the men who paid to meet him. Arsenault reportedly traveled to a number of cities as an escort, including Boston and Las Vegas. Arsenault described these trips—some lasting nine days—as glamorous.

Eventually, Arsenault's job as a University instructor 'got in the way' of the excitement of international escorting and sex work. When his contract came up for renewal at the end of one semester, he reportedly elected not to renew.

In 2000, Arsenault got a call from Zev Shalev at Alliance, regarding a reality television series, The Lofters. [3] Shalev's creation, The Lofters, exposed the daily lives of eight producers. Shalev, who envisioned a number of story lines revolving around Arsenault, hired him for the series.

Bruce Glason, a producer at PrideVision, saw Arsenault on The Lofters, and reportedly spoke to fellow producer, David Dean Portelli, about Arsenault. Portelli already knew of Arsenault and felt Arsenault's transsexuality was an important reason to include him on the 'all gay' network. The two producers reportedly were so confident that Arsenault would work on the network, they reportedly never asked him to audition. Arsenault began as a segment host for PrideVision, introducing upcoming shows. The producers eventually elevated Arsenault to co-host of The Locker Room, opposite Paul DeBoy [6]. The Locker Room, created by Josh Levy and Paul Bellini, was loosely based on John Steward's The Daily Show on the Comedy Network. The program broadcasted campy skits and factious news stories. The producers of The Locker Room created a character for Arsenault—a young transsexual version of Barbara Walters—who did investigative interviews and sports dramas. The writers produced one episode where Arsenault did a story about the gay male race horse, which was mounting other horses and subsequently was not allowed to compete in the breeder's cup. Arsenault was the co-host of The Locker Room for two seasons and ended when PrideVision suddenly and without warning ran out of funds.

Arsenault, former co-host, DeBoy, and The Locker Room visionary, Josh Levy, hooked up again in the filming of Levy's DoULike2Watch.com. [4] The entire project was filmed in three weeks. Unfortunately, the movie was never distributed in theatres.

In 2003, the producers of Showcase TV hired Arsenault to be the lead character for one season of the sexually explicit series, Kink. [5] In fact, Kink was Showcase's highest rated program in 2003.

In 2003, producers of Showcase TV hired Arsenault for a principle role in Frank Pierson's mini-series, A Soldier's Girl. [6] A Soldier's Girl is about the real life relationship between Barry Winchell and Calpernia Addams and the events that led up to Winchell's murder by fellow soldiers. Arsenault plays the role of a trendy woman. The movie was nominated for an Emmy Award.

In 2004, a casting director at Global hired Arsenault for a principle role on Train 48. [7] Train 48 is (and was) shot exclusively on a train. The program, which was broadcast five days per week, is about people who take a commuter train together; the dramas; the conflicts that they have with each other. The producers of Train 48 had Arsenault play a transsexual, named Christine, who one of the main characters falls in love with. In the script, the character, which falls in love with Christine, plays a straight man, who doesn't know Christine is transsexual.

Given Arsenault opted for a plasticized look in lieu of a natural look, Arsenault was in fact poorly cast to pull off the part of a passable transsexual.

The Global network cancelled Train 48 after two seasons. Cindy Wong, the executive producer of the series, cited a management shake up as the cause. [8]

Arsenault is one of two professional male-to-female transsexual actresses working in the film industry in Toronto, Ontario. The other is Tulsi Balram. [9] In fact Arsenault and Balram both worked on Frank Pierson's film, A Soldier's Girl. [10] Arsenault reported that he regards Balram as a personal friend. Further, Arsenault features Balram in one of his columns on Fab Magazine in an apparent scandal involving a DRAG beauty contest. [11] In T-Girl column #294 Arsenault bets that his audience will be as concerned about the entry rules of DRAG competitions as he apparently is.

Mitchel Raphael had a significant hand in Arsenault's career development. Raphael, the former reporter from the National Post, who contacted him in 1998 and is now a columnist for MacLeans Magazine, helped Arsenault land opportunities on a number of acting roles, including the role on Kink. In 2005, Raphael hired Arsenault to write a bi-weekly column for Fab Magazine, when he worked at Fab as editor. Arsenault now has over twenty eight columns to his credit. The column, T-Girl— the idea of Raphael—drew the attention of the women's studies department of York University. York University reportedly hired Arsenault to speak on sexuality.

In addition to acting and writing, Arsenault reportedly works as a professional stripper. Arsenault refers to professional stripping as his 'weekly job'. In fact, Arsenault reported that he engages male audiences at the Diamonds club in Mississauga, Ontario two nights per week.

Although Arsenault began his career in acting, Arsenault reported that he hopes to move behind the camera and establish a career in screenwriting in the future. Arsenault reported sending one of his screenplays to a producer on Los Angeles for consideration.

Arsenault reported that when he began transitioning, he had few friends to draw support from. Arsenault reported that he felt that he wasn't pretty enough for anyone to like him as 'her'. Arsenault also reported that a number of his transsexual friends tried to dissuade him from transitioning altogether, arguing that Arsenault would 'never make it' as a transsexual. Arsenault stands at six feet. Arsenault recognizes in retrospect that when his friends made those statements they did so with his interests in mind. Arsenault reportedly overcame all self doubts and proceeded with his transition. Arsenault made friends as time progressed, including Dr. Keith Loukes of the Sherbourne Clinic, who he also sees for medical services.

Arsenault is reportedly no stranger to the club scene. Arsenault is often the subject of gossip columnists. On January 31, 2006, National Post reporter, Shinan Govani, reported that Tommy Lee of Motley Crew and Arsenault hooked up (or were on the verge of hooking up) at the Ultra Supper Club on Queen Street. Tommy Lee, apparently not aware of Arsenault's transsexuality, called Arsenault 'extraordinarily beautiful'. Govani reported that Tommy Lee and Arsenault flirted. Arsenault sat on his lap. Before the relationship escalated Tommy Lee's people informed him that Arsenault was in fact a male-to-female transsexual, evidence to the effectiveness of Arsenault's forty seven procedures. Tommy Lee reportedly said at one point that he needed to go to a designated smoking room. Arsenault wanted to follow in tow. When Arsenault attempted to follow Lee, Lee's people barred Arsenault from following him any further. Arsenault reported to Govani, "My body looked incredible, if I do say so myself."

Arsenault's experiences with heterosexual men (or near experiences) are not limited to Tommy Lee. Arsenault reported in Fab Magazine he has had sex with two professional athletes, a movie star, two TV personalities, CEOs of two Fortune 500 companies, four guys who worked for the mob, a string of strippers, many male models, a bunch of body builders, loads of night clubbing suburban guys—all of them reportedly 'straight'.

* * * * *

Psychologist Ray Blanchard of the CAMH articulates the etiology underlying the She-male Phenomenon in his work on partial autogynephilia. Blanchard explains whereas autogynephilia denotes a male's propensity to be sexually aroused by the thought or image of himself with female attributes, and whereas a segment of males imagine themselves, in their sexual fantasies, as complete women, partial autogynephiles imagine themselves with a mixture of male and female anatomic features. The partially autogynephilic male will imagine himself with the secondary sex organs of a female (e.g., breasts), and the primary sex organs of the male (e.g., penis and scrotum). The combination of the male having breasts and the physical appearance of a female through hormone replacement therapy, but yet having the male genitalia intact is known in street vernacular as the chick with a dick. [12]

Blanchard found that 90% of autogynephilic males acknowledge some history of transvestism. 35% acknowledge transvestism alone, and 55% acknowledge fetishism as well as transvestism.

Males with partial autogynephilia often gravitate to pornography and sex work. She-male pornography is a distinct genre of pornography. The subjects of one genre of She-male pornography are men dressed as women, with their clothing arranged to reveal the presence of a penis and scrotum. In other genres, the subjects are naked men with breast implants or men, who show the visible effects of hormone replacement therapy or both.

She-male sex workers are a distinct class of sex worker. She-male sex workers typically advertise themselves in newspapers that carry commercial sex advertisements. She-male sex workers thinly disguise themselves as 'escorts'. The advertisements advance the image of a perfectly passable woman, who coincidentally has a penis. Examples of the copy of the advertisements include the statements, '100% passable', 'none of your friends will know' and 'seeing is believing'. The advertisements of She-male sex workers never explicitly define the audience or who they take as clients. In fact the advertisements of the she-male sex worker may appeal to males having a number of erotic interests. Blanchard found that a segment of the market for she-male sex workers were males with partial autogynephilia.

Arsenault reportedly advertises himself as 'Amber'.

Blanchard found two types of transsexuals: a homosexual type and an autogynephilic type. Blanchard makes similar distinctions between she-males, partial autogynephiles and what are known as gynemimetics. Gynemimetics, as coined by Money and Lamacz, are men who live in society as women without genital sex reassignment surgery. Gynemimetics are homosexual in object choice and they are not erotically aroused by cross-dressing. Thus, the differences between gynemimetics and partial autogynephiles parallel the differences between homosexual and autogynephilic transsexuals.

Arsenault reports he is homosexual and prefers sex with men. Following Blanchard's work on what he coins partial autogynephilia, Arsenault is then gynemimetic.

Blanchard also found there to be stark differences insofar as the age of transition between homosexual transsexuals and autogynephilic transsexuals. Blanchard found the average age of transition for a homosexual transsexual is approximately 29, whereas Blanchard found the average age of that of an autogynephilic transsexual is 40. Arsenault reportedly began to affect a gender transition at age 25.

Arsenault, who underwent countless procedures, is unquestionably beautiful in the context of a male-to-female transsexual. A number of otherwise gynandromorphophiles would find Arsenault extremely attractive. However, a male that prefers a cisgender female, like Tommy Lee, would likely not find Arsenault attractive. Gynandromorphophilia largely coexists with gynephilia, which is the sexual interest towards females. However, only 5% of gynephilic men also have gynandromorphophilia. In effect, 'straight' is not necessarily straight. Sexual conditions rarely exist in isolation.

Males, who have gynephilia (e.g., the clinical term for female-orientation) and prefer cisgender women, like Tommy Lee and the 95% of the heterosexual male population like him, do not get sexually aroused by even the most attractive and passable male-to-female transsexual. As beautiful as Arsenault is, he is not in the population of women, that non-gynandromorphophilic men, like Tommy Lee, would otherwise find attractive.

Further, the gynandromorphophilic male would not find Arsenault attractive, if he underwent a GRS procedure and no longer exuded the 'chick with a dick' stereotype. The gynandromorphophile loses interest, when the otherwise male-to-female transsexual undergoes a GRS procedure and removes what the gynandromorphophile is drawn to—lips, tits and hips with a penis. The incidence of post-operative, male-to-female transsexuals, appearing in erotic pornography is extremely low. In this case, the availability of post-operative she-male pornography follows the demand.

* * * * *

According to a 1997 study entitled, First National Survey of Transgender Violence, by GenderPAC, 59.5% of transgendered individuals claim that they have been the victim of harassment and/or violence at least once in their lifetime; 19.4% claim to have been assaulted without a weapon; 17.4% claim to have had objects thrown at them; 10.2% claim to have been assaulted with a weapon; 9% claim to have been robbed. 9.2% claim to be the victim of sexual assault. Of the perpetrators 50% are total strangers; 13.2% are an employer or co-worker.

On the surface the incidence of violence appears to be strongly correlated with class. However, the incidence of violence towards transsexuals is in fact skewed towards she-males sex workers—the most vulnerable subclass of transgender population. The perpetrators are largely men with sexual interest in transsexuals, transvestites and she-males, casting doubt whether the incidents are hate crimes.

On May 25, 1996, Victoria Day, Toronto, Ontario, which has one of the lowest homicide rates per capita in North America, witnessed one of the worst serial murders of transgender individuals in the nineties. In a two hour killing spree Marcello Palma shot to death Shawn Keegan (a transvestite prostitute), Deanna Wilkinson (a transsexual prostitute) and Brenda Ludgate (a female prostitute). All of the victims were shot in the back of the head with a .357 magnum filled with hollow-point bullets.

Palma, who is the thirty-four-year old father of one child, was under a psychiatrist's care at the time of the killings, but still had six restricted weapon certificates and owned six firearms. During the therapy sessions Palma discussed having marital problems and admitted to having sex with prostitutes, transvestites and homosexuals.

According to his own testimony, in the middle of a thunderstorm as the rest of the city was celebrating Victoria Day festivities (which included firecrackers) Palma loaded his Sturm Ruger .357 with five illegal hollow-point bullets, took the handgun and a knife and drove off in his red pick-up truck. At approximately 11:00 p.m. he picked up Ludgate and drove to the back of a lighting company where he asked her to perform oral sex for him. Before the act, he changed his mind and asked her to get out of the truck. When she wouldn't leave, he hit her, pulled her from the vehicle and shot her in the back of the head.

About forty minutes after Palma shot Ludgate, he picked up Keegan, dressed as a woman. A struggle ensued at the top of a stairwell of an apartment building on the corner of Carlton Street and Homewood Avenue. Palma pushed Keegan down the stairs and shot him in the back of the head. As he put up his umbrella he observed Keegan "kind of get up" and shot him again.

Minutes later Palma crossed paths with his third victim, Wilkinson, as he walked along Homewood Avenue. Moments later a witness heard a man yell, "bastard", and then a gun shot. Wilkinson's body was not discovered until 9:00 a.m. the next morning. Wilkinson was also shot in the back of head.

Palma's lawyers argued that Palma suffered from a bi-polar disorder, and was not criminally responsible for the murders. However, the Crown Attorney was able to demonstrate that Palma had premeditated the murders, conferring with a number of individuals over a long period of time that he intended to take care of the "trash" on Homewood Avenue. Palma was referring to the transsexual sex workers, who operate on Homewood Avenue. Homewood Avenue is in the heart of the gay district of Toronto, Ontario.

Ultimately, Palma was found guilty of the three murders and his defense that he was not criminally responsible at the night of the murders was struck down.

A hate crime is one where the perpetrator selects his victim completely on the immutable characteristics of the individual, in this case transsexual; however, the crime must cross subclasses. Palma, otherwise, paid the people, he victimized, for sex acts. Given Palma's proclivities towards transvestites, transsexuals and she-males, given gynandromorphophilia—the sexual interest in transvestites, transsexuals and she-males in men—is related to partial autogynephilia and gynemimicphilia, Palma's act does not qualify as a hate crime. Palma preyed on members of his own community in relation to his underlying disorder and the underlying disorders of the people he victimized.

Palma heinous acts are typical of those of other incidents of violence towards transsexual sex workers. In fact the nature of the violence tends to be severe and in many cases often involves mutilation of and around the genital areas. Tasha Dunn, a 42 year-old sex worker, was found bludgeoned to death in a Tampa parking lot on Valentine's Day, February 14, 1998. Christian Paige, a 24 year-old male-to-female transsexual and native of Nashville, Tennessee, was found dead in his apartment on March 22, 1996. Page had been savagely beaten in the head, strangled, and stabbed in the chest approximately 15 to 24 times. Several of the stab wounds were located on and around his breasts. Before fleeing, his assailant set fire to the apartment, in an apparent attempt to destroy the evidence. Paige reportedly competed in a Miss Continental contest in Chicago, Illinois just prior to his murder.

Mary S., a transsexual sex worker and former programmer/analyst, was found in the trunk of a car in a river in 1993. Although the former computer programmer/analyst had multiple lacerations to his body it was not clear whether he died from drowning or from the multiple stab wounds. Mary's odyssey into sex work reportedly began when he began to transition on the job as a computer programmer/analyst position in Boston. As he began transition his employer summarily fired him.

* * * * *

The laws regulating prostitution vary greatly from jurisdiction to jurisdiction. In some jurisdictions prostitution—generally defined as the provision of sexual services for money—is itself illegal; in others the act of prostitution is not illegal in itself, but many associated activities are unlawful. In England, for instance, the law does not prohibit prostitution itself but does prohibit soliciting for prostitution in a public place, living on the earnings of prostitution, exercising control over prostitutes, or keeping a brothel (any premises where two or more prostitutes are employed).

In some jurisdictions, notably in the U.S. state of Nevada, prostitution is lawful and practiced openly subject only to health and related controls. Sex workers (non-transsexuals and transsexuals alike) are required to undergo medical examinations on a regular basis. If they do not pass health standards, they lose their privilege to work in the industry.

As society largely regards sex work as illegal; society so too regards the individuals who participate in the industry and particularly the sex workers themselves as immoral and subject to nefarious influences. Sex workers are stereotypically lumped in with pimps and hard core drug traffickers. Ironically, very few who harbor these opinions have ever met a sex worker in person. In fact, only a handful of individuals in society regularly engage sex workers. The people who do, regard sex workers (and in particular the ones they engage) in a positive light.

"They worship sex workers." Christina Strang says. Christina Strang is the coordinator of the volunteer program, Meal Trans, which provides support and counseling to under-privileged, transsexual sex workers. Prior to being the coordinator of Meal Trans, Christina worked as a sex worker "herself" from the age of 18 to the age of 23. "The vast majority of men who regularly engage sex workers treat them very well. The degree to which they treat them depends on their own economic situation. If the John is wealthy, he is more likely to be generous with the girl that he lets take care of him."

The proportion of transsexual sex workers as compared to the population at large is significant. It is thought that there are approximately twenty five thousand male-to-female transsexuals in North America. If this is correct, transsexuals make up a very small percentage of the population. (~0.01%)

Blanchard and Collins (Clarke Institute of Psychiatry) examined 10 consecutive issues of a free weekly Toronto news and entertainment magazine [1993] with an audited circulation of approximately 95,000 copies. The average number of advertisements that included blunt or thinly disguised prostitution advertisements was 395 of which 19.6 (~5%) were placed by individuals who described themselves as transvestites, transsexuals or she-males. This suggests that the number of males or females who engage transgendered escorts encourage at least 19 of these individuals to advertise every week in that one periodical. All else being equal this suggests that 5% of the supply of sex workers in the industry is in fact transsexual (or some variation of transgender males). Thus, the number of transsexuals operating in the sex industry is disproportionally high as compared to society at large.

Why do transsexuals get involved in sex work? The answer is simple: economics, control and freedom. Sex workers can earn significant sums of money. "Sex work compares with any job out there in that it pays as well." Strang says. "The amount of money a sex worker can make, varies from person to person."

Whereas there are a number of other vocations that pay equally well as sex work, sex work offers one thing that these conventional jobs don't. "My business. My control." Strang says.

"The sex worker in this case doesn't have to answer to a boss per se. They are in effect self-employed. Like any other business they have to hustle clients to engage their services. The successful ones the ones that are able to make a career of sex work, are able to turn the one time 'Johns' into repeat customers, many of which continue to engage the sex worker for years."

Apart from the sheer economics a number of transsexuals regard sex work as much more self-fulfilling than say more conventional jobs. "There is a unique culture associated with the trade. These people know each other on an intimate basis. Many of them, who could earn more money elsewhere, remain in the trade due to their love of the culture. There's nothing like it."

Contrary to popular belief that transsexuals are the victims of marginalization and turn to sex work as a last resort, the vast majority of transsexuals who enter the industry have work experience as a female and are capable of finding employment or discovering other ways of generating an income.

"Whereas about half of the sex workers are transient and move on to pursue other employment, the other half of the transsexuals in the industry are career sex workers, despite their ability to pass [as females in society]." Strang says. "Transsexual sex workers are much more proud to be transsexuals than other transsexuals. They are committed to the community and the culture associated with it." Another myth surrounding the transsexual sex industry has to do with the profile of the clients who engage sex work services. It has often been thought that the vast majority of males that engage transsexual sex workers are homosexual. In fact, the opposite is true. The vast majority of men who engage transsexual sex workers are in fact heterosexual (meaning that they have a heterosexual history). Dr. Blanchard of the Clarke

Institute coined the term, gynandromorphophiles, to designate males with distinct interest in feminized males, including men wearing women's clothing and men that have altered the contours of their body to a more feminine appearance by either surgical or hormonal means, but leave their male genitalia intact.

"They like a woman with a penis. In most cases these men have wives or girlfriends, but to them they are experimenting. In other cases where these men already have transsexual partners they are able to get something from the sex worker that they can't get from their transsexual partner." Strang says.

The sex industry has a number of occupational hazards that the transsexual sex worker must deal with. Perhaps the biggest problem that these professional transsexual sex workers face is society's attitudes regarding the sex worker in general.

"The clients do not see sex workers as people. That's the biggest problem." Strang says. "These guys want to have sex with a woman with a penis. However, in most cases, they don't want to date them. They already have wives or girlfriends. If society's attitude toward sex work were different, a number of girls would have a more healthy self-esteem. The way things are it's easy for a transsexual sex worker to get down on herself. It's easy for her to get down on her body."

As reported above Richard Adamson, a.k.a. Rikki Dreamy Sweet, died of drug overdose in the early morning of Saturday, June 2, 2001 at his apartment in Toronto, Ontario. Sources close to Adamson indicate that he had been concerned about his finances. These same sources say that he was taking on a number of 'Johns' without screening them properly just to pay rent. One source used the word, 'desperate,' to describe Adamson's actions. However, the sources claim that Adamson concern was not such that he talked about the possibility of suicide. It's hard to imagine a situation where someone is failing to pay his expenses and not have it affect their self-esteem and ultimately their mood.

"I don't know what guys want from me." Adamson is on record for saying at one point prior to his death.

The attitudes of the main stream have ripple effects that the sex workers and other marginalized groups finally bear. The worst ripple effect from this negative attitude is violence against sex workers and specifically violence against transsexual sex workers.

Nearly every sex worker has had at least one experience of being assaulted. In March, 1997 Gabriel Gladeu was assaulted by a first-time John. After Gladeu performed oral sex on the individual, he asked to be paid. He refused, citing he was not satisfied with his work. A scuffle ensued. The John reportedly shoved Gladeu through a plate glass window. Gladeu has a number of deep scars across his wrists as a result.

The 519 Community centre and Meal Trans have amassed a large database of individuals that the members of the community are aware of as potentially threatening. These are individuals that transsexual sex worker report to the centre as being violent or having the propensity of being violent. They are linked in with other community support groups and share information.

The other area of concern is health. Sex work involves physical contact. When you go visit your dentist, chances are he/she wears rubber gloves when rendering his/her services. He/she is wearing gloves so to not transmit disease that could affect him as well as you. Sex workers have their own tools to prevent the spread of disease. They have condoms. However, that will not prevent sex workers from contracting other diseases that can pass through the saliva. Sex workers are vulnerable to contracting Hepatitus C, a liver disease. Hepatitus C can be fatal if not treated properly. Richard Adamson, (a.k.a., Rikki Dreamy Sweet) suffered from this terrible illness during the last year that he was alive.

In addition to these hazards, transsexual sex workers have to take a myriad of medications to transform their bodies into that which resembles a female. All prescription medications have their side effects—many adverse ones. Further, a number of transsexuals purchase female hormones on the black market and medically treat themselves. It is not completely uncommon that sex workers suffer ill-effects from feminizing medications. For instance, estrogen and progesterone, affect one's mood. Estrogen alone can cause depression.

Worse still, the chance exists of potentially fatal side-effects whether an individual is under a doctor's care or not. _Transgender Care_ (Temple University Press, 1997) has this to say: "Complications for transgender women with biological risk factors or for those who have had excessive estrogen therapy include [...] thromboembolism (blood clots in the legs, lungs, eyes, brain - as in strokes - or other organs) as well as others more long-term, such as breast cancer and liver disease."

I interviewed two transsexual sex workers, who developed a blood clot from synthetic estrogen. One developed a blood clot in one of his testicles and later had to have the one testicle surgically removed. The treating doctor in his case advised him not to restart hormone replacement therapy.

Given the fact that sex work is illegal (at least in the vast majority of states in the U.S. and Canadian Provinces, including Ontario), sex workers who work the streets are subject to arrest. An arrest will effectively take the sex worker out of commission for weeks at a time. Although this is not an occupational hazard per se, having a criminal record is definitely a drawback for sex workers. In the time the sex worker is in the correctional facilities awaiting court appearances, he may lose his regular clients who will eventually seek the service from other competitors.

The quality of services that reach out to sex workers varies from municipality to municipality. In Toronto, Ontario, Meal Trans, which operates out of the 519 community centre in the heart of Toronto's gay community, offers counseling and support services to street active transsexuals. The staff at the 519 Community Centre is a volunteer one. And the program operates largely on charitable donations. To combat violence against transsexual sex workers, the Toronto rape center through Meal Trans and the 519 community centre teaches young transsexuals how to defend themselves in case of sudden attack.

references

[1] GENDER PRONOUNS

In the book I utilize male appropriate pronouns to describe what are known as male-to-female transsexuals and female appropriate pronouns to describe what are known as female-to-male transsexuals. The reason is to provide clarity with respect to the person's biological anatomy.

It is legally possible in North America for individuals to change their sex code. The requirements differ from jurisdiction to jurisdiction. It is generally politically correct to respect transsexuals by using pronouns that are consistent with the gender they identify themselves to be. However, a person's legal sex and anatomical sex are not necessarily the same.

Despite the advances in medical science, no technique or procedure exists to change an individual's anatomical sex. Medical science has had some success altering a person's outward gender appearance through the introduction of synthetic hormones. Medical science has also had some success in cosmetically fabricating the anatomy of a penis into what looks like a vagina. Medical science has also had some success in cosmetically fabricating a male penis by way of what are little more than stents and skin grafts. These medical successes, however, are extremely limited.

Further, the medical advances fall far short of satisfying the basic definitions of what is female of a species and what is male.

To illustrate the point in the case of a male-to-female transsexual, when he undergoes a penectomy and orchiectomy, he disables his ability to fertilize, which is a critical function of the male species, but he does not advance his ability to conceive or bear offspring, which are critical functions of the female of the species. In the case of a female-to-male transsexual, undergoing testosterone disables her ability to conceive, however, the attachment of a stent does not advance her ability to fertilize, which is a critical function of the male of the species.

Even the most 'passable' transsexuals have moments when they fail to pass. {Maxine Peterson, formerly Leonard H. Clemensen, M.A., CAMH, 2002]

[3] U8TV: THE LOFTERS

U8TV: The Lofters is a reality television show released January 15, 2001.
more information .
visit the site: IMDB U8TV: The Lofters
http://www.imdb.com/title/tt0274296/

[4] DOULIKE2WATCH.COM

Jose Levy's DoULike2Watch.com was produced in 2003.
more information
visit the site: IMDB DoULike2Watch.com
http://www.imdb.com/title/tt0414949/

[5] SHOW CASE'S KINK

Show Case's Kink is a cable television show that has been operating for five seasons.
more information
visit the site: Show Case's Kink
http://www.showcase.ca/kink/

[6] PIERSON'S SOLDIER'S GIRL

Frank Pierson's directed the film, Soldier's Girl, in 2003. The film featured Lee Pace as Calpernia Addams. The cast also includes Toronto, Ontario based actors Tulsi Balram and Nina Arsenault in a principle role.
more information
visit the site: IMDB Soldier's Girl
http://www.imdb.com/title/tt0324013/

[7] GLOBAL'S TRAIN 48

The Global Television Network broadcast the television show, Train 48, in 2003. Steve Levitan and Paul Bronfman of Protocol Entertainment produced the show that was filmed entirely on a commuter train. The cast includes Nina Arsenault in a principal role.
more information
visit the site: Global's Train 48
http://www.protocolent.com/train48/

[8] CBC REPORTS GLOBAL CANCELS TRAIN 48

On June 2, 2005 CBC reported that the Global Television Network cancelled the television series, Train 48.
more information
visit the site: CBC article, Global Cancels Train 48
http://www.cbc.ca/arts/story/2005/06/02/train050602.html

[9] INTERVIEW WITH NINA ARSENAULT, DATED APRIL 6, 2004

I conducted a number of interviews with Nina Arsenault, a.k.a., Rodney Arsenault, a.k.a., Amber, including one on April 6, 2004. The audio material is available via the links below.
more information
Interview April 6, 2004
http://www.transgression.com/Assets/Downloads/interview%20arsenault,%20dated%2020040406,%20VBR%20low.mp3

[10] PIERSON'S SOLDIER'S GIRL

See [6]

[11] NINA ARSENAULT TGIRL COLUMN FOR FAB MAGAZINE

Nina Arsenault wrote a column for Mitchel Raphael of Fab Magazine, Tgirl.
tgirl's archive
Fab Magazine, T-girl issue #276
http://www.fabmagazine.com/tgirl/archive/276/index.html
Fab Magazine, T-girl issue #294
http://www.fabmagazine.com/tgirl/archive/294.html

[12] THE BLANCHARD ARTICLE, THE SHE-MALE PHENOMENON AND THE CONCEPT OF PARTIAL AUTOHYNEPHILIA

Ray Blanchard published the article, The She-Male Phenomenon and the Concept of Partial Autogynephilia, to forward his work on the study of what he refers to as partial autogynephilia.
The She-Male Phenomenon and the Concept of Partial Autogynephilia; Journal of Sex & Marital Therapy, Vol. 19, Spring 1993, Brunner/Maxel, Inc..
more information
visit the site: Abstracts of Relevant Articles by Blanchard
http://www.autogynephilia.org/Abstracts.htm

CHAPTER 09

The Missing Concept of Autoandrophilia

It is January 28, 1998, at 9:15 p.m. at the lower floor bar area of Pimplett's English Tavern. Pimplett English Tavern is the creation of Geoffrey Pimblett, a 52-year-old English, cross-dresser from Wales. In fact, Pimblett has two ongoing business concerns, including the restaurant on the south side of Gerrard Street East and a Bed & Breakfast on the north side of Gerrard Street East. The two buildings, which are located in the gay district of Toronto, Ontario are within two hundred yards of each other. Pimblett's English Tavern, like the Take a Walk on the Wildside Fetish Bed & Breakfast, is well-known to largely straight, closeted, cross-dressers across North America and Europe. 'Paul' is at the bar, alone by herself.

Paul (not her real name) is a public, female cross-dresser. If Paul didn't introduce herself as Paul, you wouldn't know that anything were out-of-the-ordinary. If Paul didn't introduce herself as Paul, you may mistake Paul for a homosexual women, but not necessarily a butch dyke lesbian. Paul is a 31-year-old heterosexual woman, who has a history with men. In any event Paul reported that she is not involved with anyone at the moment.

Paul is out numbered in the lower floor area adjacent to a bar by her male counterparts twelve-to-one.

Paul is decked out in a double-breasted, Olive-colored suit with the sets of buttons to the right—which is appropriate to men's clothing—flashy suspenders, dress socks, wing-tipped shoes, which appear over-sized for "Paul's" small frame— Paul is only five foot six inches tall. "Paul's" wing-tipped shoes appear to be a men's size 8-and-a-half. Paul is also slight. "Paul's" white dress shirt hangs loosely on her body. Paul has no evidence of breasts pressing against her dress shirt. It is not clear, if Paul has undergone a mastectomy, or whether Paul is simply small-breasted and has her breasts wrapped tightly to her body. Finally, Paul has the hair style of a short barber cut in men. Paul has left nothing on the table.

Paul reported that she has not yet started hormone replacement therapy, however, Paul reports that she does plan to do so "shortly."

If Paul progresses and undergoes a phalloplasty her chances of experiencing orgasm similar to that of a male with the born anatomy of a penis is nil. Jamison Green correctly reports the state of transsexual surgical procedure and their limitations in "his" article "Getting Real About Genital Surgery". Paul as a female-to-male transsexual will likely find it equally difficult finding a male partner as a post-operative male-to-female transsexual finding a natal female partner. Paul may be able to draw sexual satisfaction from exercising a behavioral addiction kicked off by the consideration of being a man, but she will likely never progress beyond asexual practices. If Paul progresses and continues to take synthetic testosterone, she will also face enormous health challenges and will likely die at a younger age than if she never moved forward with the gender transition.

* * * * *

Given the small incidence of female transsexualism, given female-to-male transsexuals more quickly assimilate in society as visible men, it is rare to witness their life narratives. The masculinizing effects of synthetic testosterone allow females to assimilate into society as visible men more readily than males attempting to blend as females. Female-to-male transsexuals, who undergo hormone replacement therapy, are in fact visible for only a brief time. After that they emerge as often diminutive, balding men with undersized hands and feet.

Over a three month period starting in June, 2004 Hana Gartner and the CBC Fifth Estate captured a rare glimpse into the gender transition of a young female-to-male transsexual, in the segment, Becoming Ayden. Ayden—Adina Scheim—is a 17-year-old girl from Toronto, Ontario. Adina is the product of a broken home. Her father, Phil Scheim, a conservative Rabbi, and her mother, Lori Scheim, split when Adina was 13. The Scheims gave conflicting evidence about their daughter. The father reported Adina as otherwise a normal girl, whose problems began at age 13.

"She's a brilliant girl. She could have pursued any career she wanted. But now I think she's really limiting herself to a marginal element in society and that's hurtful." Phil Scheim says.

Lori, Adina's mother, on the other hand reported that Adina was always 'conflicted'. In the program Adina is just beginning to undergo Hormone Replacement Therapy under the direction of Leslie Shanks, the Medical Director of the Sherbourne Health Centre. The Sherbourne Health Centre is a third wave provider of health services to transsexuals. Shanks is of the belief any delay in Adina's gender transition would be harmful and costly. Whereas Adina's parents want Adina to suspend a gender transition until after she completes University, Adina wants to move forward with her gender transition.

In one segment of the program Hana Gartner and her camera crew follow Scheim and Evan Smith, a female-to-male transsexual friend, to Seattle, Washington as the two 'boys' attend a Female-to-Male (FTM) conference. Although the conference has a community side, the conference also provides information on the availability of surgical procedures geared towards female-to-male transsexuals. Effectively, Scheim and Smith are out shopping for surgical procedures. In the consideration of one particular surgical procedure Scheim visibly appears giddy on camera.

Adina is otherwise straight and as Ayden she identifies as being queer. Smith is reportedly also straight.

There is no evidence that Shanks, who is Adina's treating physician, ever considered Adina's symptoms as anything other than a gender identity disorder. There is no evidence that Shanks ever considered Adina to have a behavioral addition kicked off by a sexual response of the consideration of a gender transition. Further, there is no evidence that Shanks ever considered any treatment alternatives to hormone replacement therapy, leading to invasive surgeries. Given Adina's age, given Shanks has no consent from Adina's parents, Shanks' approach to treating Adina is ill-advised.

After the show's airing, Scheim and Smith were critical of the Hana Gartner and the Fifth Estate for the manner in which they portray Scheim. They were also critical of the manner in which they covered Dr. Pierre Brassard's attitudes with respect to the effectiveness of gender reassignment surgical (GRS) procedures. In the segment Brassard is evasive. In response to the question of whether the surgery is in the health interest of his patients, Brassard said he didn't know. The implication is despite earning millions of dollars performing surgical procedures on transsexual patients that he is unconvinced what he does is of any value or whether what he does is in his patients' health interests. However, it is not his job to know. Dr. Brassard's job is not to screen candidates for GRS procedures. (In fact many lie to their respective psychologist and arrive at the Montreal clinic having never lived a day as a member of the opposite sex.) Rather, Brassard fully relies on the psychiatric community to properly diagnose and screen candidates for GRS procedures. If the psychiatric community fails to properly screen a candidate, Dr. Brassard is not liable. Brassard's only concern is to perform the procedure to the best of his surgical abilities. Brassard, himself, may or may not believe in GRS procedures. It is not clear. What is clear is if the body of experts concluded that a GRS procedure is not in the health interests of a sexually developed patients, Brassard (at age 50 or thereabouts) is not likely to seek another means of earning money.

The CBC Fifth Estate's 'Becoming Ayden' accentuates the journalistic challenges of delving into a politically charged topic. Gartner does a superb job at resisting the role of a gender advocate while bringing Scheim to face the rationality of a gender transition at such an early age—a contentious issue with her parents. At one point Scheim walks off the set in disgust, believing Gartner has taken her parents' side.

* * * * *

The incidence of transsexualism in females is low—the male-to-female ratio of pathological transsexualism is 3 to 1. Psychiatrists and psychologists know very little about transsexualism in women like Paul. In fact, even psychologist, Dr. Ray Blanchard of the Centre for Addiction and Mental Health (CAMH), who has dealt with over three hundred case files of women, seeking surgical procedures to change their basic sexual anatomy, is silent on the matter, leaving only Gender Identity Disorder to speak for their life narratives. A Gender Identity Disorder (GID) is again the persistent discomfort with one own sex / gender role in social setting coupled with the wish of being a member of the opposite sex. No one in psychology or psychiatry is on record for suggesting the female-equivalent of autogynephilia—autoandrophilia—exists in females.

There are a number of factors to account for the dramatically low incidence of the transsexualism in females, including heredity and addiction and societal influences. Medical researchers have shown that hereditary plays a significant role in addiction. The gene is passed from father to child. The propensity for the gene associated with addiction to pass to male offspring is higher than female offspring. It is unknown how big the extra propensity is, but the children of parents with alcohol addictions, for example, must keep in mind that they have a greater chance to develop problems with alcohol and other behavioral addictions, including the behavioral addiction underlying transsexualism.

The higher propensity of the gene to pass from father to son rather than from father to daughter could explain why transsexualism as a behavioral addiction has a far greater incidence in males than females. The ratio is also significant. The male-to-female ratio of pathological sex addiction is 3 to 1 (3:1)—virtually the same male-to-female ratio of transsexualism. Further, the male-to-female ratio of pathological gambling addiction is 2.8 to 1—again, virtually the same male-to-female ratio of transsexualism. [2] The male-to-female ratio of pathological transsexualism is 3:1. [3]

In 1920 (over eighty years ago) with the passage of the nineteenth amendment to the constitution of the United States, women won the right not only to vote, but also to transgress gender norms. The net result was the gradual end of gender oppression and huge gains in the productivity of the work force. As women were allowed to transgress gender norms as women were taking on greater responsibilities in the workplace, the incidence of paraphilic tendencies to become aroused by the thought of being or turning into a man, largely evaporated, but not completely.

In any event the implication of the low incidence of autoandrophilia in females has implications towards transsexuality in men. If men were allowed to openly transgress gender norms in society, the incidence in the paraphilic tendency to become aroused by the thought of being or turning into a woman, would similarly evaporate like it largely did for females. If men had no behavioral addictions kicked off by a paraphilic response to the consideration of being female, there would not only be a smaller incidence of transsexualism in men, but also the probability of a male-to-female transsexual, having an affected gender identity disorder, to benefit from genital surgery would increase dramatically. (See Chapter 11, The State of Gender Reassignment Surgery in the Twenty First Century).

*　*　*　*　*

The so-called gender continuum includes two fixed positions, male and female. The forces that shape a male from a female occur entirely in the mother's womb. When the child begins to sexually develop, at either age 5 or at age 11, the child has a sexual identity that is immutable. The people, who gravitate to identify as somewhere along the gender continuum, practice nonconformity to differentiate themselves from the crowd. A person along the gender continuum generally likes to think of themselves as somehow 'special' and they are, but not for the reason they advertise.

references

[1] THE REPORTS OF BLANCHARD BEING, HIMSELF, TRANSSEXUAL

Patricia Aldridge, who owns the self-described fetish boutique, Take a Walk on the Wildside, reported that Ray Blanchard is himself, a female-to-male transsexual. I could not confirm this claim.

[2] WENZEL, HANNE GRO, DAHL, ALV A.: FEMALE PATHOLOGICAL GAMBLERS—A CRITICAL REVIEW OF THE CLINIC FINDINGS

WENZEL, HANNE GRO, DAHL, ALV A.: (2008) Female Pathological Gamblers—A Critical Review of the Clinic Findings. International Journal of Mental Health & Addiction, January 12, 2008, ISSN: 1557-1874 (Print) 1557-1882 (Online); DOI: 10.1007/s11469-008-9174-0.

[3] EKLUND, P.L, GOOREN, L.J., BEZEMER, P.D.: PREVALENCE OF TRANSSEXUALISM IN THE NETHERLANDS

EKLUND, P.L, GOOREN, L.J., BEZEMER, P.D.: (1988) Prevalence of Transsexualism in the Netherlands. British Journal of Psychiatry, 152: 638-640 (1988)

CHAPTER 10

The DRAG King Survival Guide

It is 6:00 a.m. on Friday, October 27, 2003. Enza "Supermodel" Anderson arrives at Woody's located at 465 Church Street in downtown Toronto, Ontario. Woody's—which is adjacent to Sailor at 467 Church Street—is a bar in the heart of the city's gay district. Just last night, Woody's celebrated the re-election of incumbent city councilor, Kyle Rae, in municipal elections of ward 27—Toronto Centre-Rosedale. Hundreds of supporters—mainly Socialist/New Democrats—were at Woody's celebrating Rae's re-election. "Supermodel" Anderson, who works at Woody's as a janitor, arrives to clean up after the huge festivities.

"Supermodel" Anderson is a seasoned DRAG performer. However, during his day job as a cleaner at Woody's, Anderson is dressed down, wearing a sweat shirt, a pair of track pants and a touque over his head. "Supermodel" Anderson is otherwise bald and shaves his scalp. Anderson reports that it is easier to wear wigs with a shaved head.

"Supermodel" Anderson was born Vince Difazio in 1964. Anderson grew up in Toronto, Ontario in what is known as the Jane Finch corridor. The neighborhood today has a high crime rate.

"Supermodel" Anderson attended York University to study geography. However, Anderson dropped out due to a general lack of interest in school. Later, after a one year hiatus, Anderson attended Seneca College where he graduated with a Civil Engineering and Technologies diploma in 1988.

After graduating from Seneca College, Anderson went to work as the quality control supervisor of a concrete pipe-manufacturing plant. He reportedly hated his job. Five years later, when the construction industry went into a downturn, the engineering firm laid him off. Anderson reportedly felt relief from being let go. While Anderson was out-of-work, his DRAG career reportedly "took off". He began to perform on a regular basis.

DRAG is not a highly sought after form of entertainment, even in a city as large as Toronto, Ontario. The vast majority of DRAG queen performers perform on stage 'sans gratis'. The top performers earn approximately two hundred to three hundred dollars (CDN$200 -

CDN $300) per evening. The performers largely lip sync to the songs of vintage Motown performers. They tell jokes in between songs.

Given the limited number of venues, Toronto, Ontario has approximately eight to ten professional DRAG queen performers at any one time. Few earn any appreciable money from performing DRAG. Chris Edwards is one exception. Edwards reportedly earned enough money through DRAG to undergo a breast augmentation procedure. Michelle Ross, a Diana Ross look-a-like, is another exception. Ross also earns a consistent and predictable income through DRAG. Ross has reportedly not undergone any feminizing procedures.

DRAG Kings are even less in demand. Toronto, Ontario does not support one professional DRAG King performer.

The vast majority of DRAG venues are effective gay bars. They are physically located in the heart of Toronto, Ontario's gay district on Church Street. However, the El Convento Ricos, which is on College Street, features DRAG of a different nature. The Ricos integrates DRAG in Salsa dance.

In 1995, Anderson got a job at a Bonival hair salon on Yonge Street, handing out flyers for the salon to pedestrians in DRAG. To supplement his income Anderson signed up for hospitality gigs, hosting organized events.

On June 18, 2000, or thereabouts a photographer from the Sun Media snapped a picture of Anderson kissing mayor, Mel Lastman, as Lastman attended a flag raising ceremony to mark Toronto Gay Pride week. The picture that below read "Mel has a Gay Ol' Time" lifted Anderson into the public eye. After that one event, Anderson, who reveled in front of the camera, never looked back.

"Supermodel" Anderson reported that as a man, he is a wimp. However, when he assumes his "Enza" alter ego, he reportedly feels empowered.

"Supermodel" Anderson reported that he began to cross-dress at age 14. Anderson, who has no siblings, reported that at age 14 he got into his mother's clothing and shoes. Anderson has a history of sexual relations with both men and women, however, Anderson reports that he prefers sex with men. Anderson engages in gender role play where he is the woman and the other party, usually a man, plays the role of the man. Anderson also engages in incestuous role play where he directs the third party to play the role of his father and he is the transsexual daughter, who needs a gender reassignment surgical procedure. In the ritual Anderson asks the man to bring him a Barbie doll.

"Supermodel" Anderson and Rae are more related than the scene suggests. Anderson opposed Rae for the city councilor's position. In fact Anderson—who eagerly watched the results at the home of a friend—placed second in the contest. However, it is generally believed that "Supermodel" Anderson was not a real threat to unseat Rae as city councilor.

"Supermodel" Anderson is not new to politics. In 2000, Anderson ran for Mayor of Toronto, Ontario. That year incumbent, mayor Mel Lastman ran away with the contest with 483,277 votes. However, Anderson placed a respectable third with 13,585 votes. Tooker Gomberg followed Lastman with 51,111.

In 2002 "Supermodel" Anderson ran for the leadership of the Alliance Party, following the resignation of the party's embattled leader, Stockwell Day. Anderson ran on a platform of bringing broad inclusion into the otherwise conservative, Western-based party.

"The reason I am campaigning for the leadership position of the Alliance party is to make a difference; to help out in a positive way. My political aspirations are not as large as a number of people in the media have characterized them to be." Anderson says.

"In addition, I hope to gain a greater knowledge of politics in general. Not only how politics works, but also how to deal with people and the public. Political skills are very important in day-to-day life. When you are engaged with other individuals, you have to assess where people are coming from, what areas are of concern to them; you have be able to give and take; and you have to be open to other people's ideas. That's what negotiation is all about."

During a January 6, 2002 press conference, "Supermodel" Anderson gave brief answers to the questions from the media. When asked what he would hypothetically do to strengthen the value of the Canadian dollar if elected Prime Minister, Enza replied that he would "sleep" with the Finance Ministers of the other G7 countries.

Ian Ross—Anderson's campaign manager—worked out a deal with his "connections" in the Progressive Conservative Party—a rival of the Alliance party on the right—to fund Anderson's Alliance party candidacy. As a first step the candidates for the leadership of the Alliance party must ante up a $25,000.00 deposit. Ross reported that his connections in the Progressive Conservative party (the Tories) had pledged that amount with no strings attached. In effect, the Tory PC party backers under Joe Clark's leadership agreed to underwrite the first step in Anderson's campaign. However, the $25,000.00 candidate deposit would only mark the beginning of the campaign expenses. There would be more—a lot more to come. Ross reported that "Supermodel" Anderson would require an additional $100,000.00 to run a proper national campaign. That would have otherwise moved "Supermodel" Anderson into the big time, politically. When he ran for mayor, "Supermodel" Anderson spent a mere $300.00, something he had pointed towards as a demonstration that he is fiscally responsible. Further, Anderson still has his only sign from that campaign. It is made from plastic and is recyclable, which makes Anderson environmentally conscious, another strong point in his favour.

To raise the necessary funds, "Supermodel" Anderson asked members of the GLBT community to not just spend $10.00 for an Alliance membership so they can vote for him, but also to come up with an extra $10.00 or $15.00 to help finance his campaign. GLBT groups generally have little to do with conservative parties, like the Alliance party and the Progressive Conservative party otherwise. Anderson stressed that the extra contribution will only be used in his campaign and not go towards the Alliance party. Ross organized a number of fund raisers, including one at Woody's on January 10, 2002, and one at the Buddies in Bad Time Theatre on January 18, 2002.

Ian Ross also chaperoned "Supermodel" Anderson to a number of "all candidate" debates, including one at Chatham, Ontario. All candidates debates are essentially open forums where candidates have the opportunity to address delegates for their support. Anderson spoke at the Chatham, Ontario event. Unfortunately, Anderson was visibly nervous and was not well received by the otherwise conservative delegates. A number of the delegates regarded "Supermodel" Anderson as a 'joke' and were rude to him. However, a number applauded him for his efforts.

In the week leading up to the deadline for candidate's deposits, "Supermodel' Anderson's Tory backers reportedly introduced a caveat to the deal behind the deposit, demanding that Anderson follow a script of a loud and obnoxious candidate throughout the leadership campaign. Faced with the prospect of public humiliation, "Supermodel" Anderson and Ross refused the funds dangled in front of them by their prospective Tory's backers and on January 31, 2002, Anderson dropped out of the race, falling approximately CDN$17,000.00 short of the CDN$25,000.00 deposit to continue.

The Alliance party of Canada ultimately chose Stephen Harper as party leader. In 2004, the Alliance party and the Progressive Conservative party merged to become the New Conservative party of Canada with Stephen Harper again at the helm.

Carlos Valencia chronicles "Supermodel" Anderson's ill conceived campaign to crack the otherwise ultra conservative Alliance party in his documentary, A Man in a Dress. [1]

Valencia's camera catches "Supermodel" Anderson at his lowest depths —deceived and deserted by his Tory backers, who had promised him the funds necessary to challenge Stephen Harper and Diane Ablonczy, but only on the caveat that he humiliate himself and the Alliance Party in the process.

By 10:00 a.m. in the morning of October 27, 2003, Anderson finishes his shift at Woody, including the toilets that Rae and the other members of his celebration party had urinated in to the night before. "Supermodel" Anderson heads back to his studio apartment located on Wellington Street that he rents for $585.00 or thereabouts. The walls of his apartment are dark from the purple paint, which has not changed in years. There he falls asleep. Anderson reportedly naps after his morning shift.

"Supermodel" Anderson, who gave over one hundred interviews would, however, not be denied. One year later, in 2004, Anderson got a job at the Toronto publication, Metro, covering local entertainment, celebrity gossip and reader profiles and quit his minimum wage janitorial job at Woody's. At Metro Anderson reportedly earns $375.00 per week for two columns. That year "Supermodel" Anderson also quit politics, DRAG, and began hormone replacement therapy.

references

[1] CARLOS VALENCIA'S DOCUMENTARY—THE MAN IN A DRESS

Carlos Valencia and Julie Giles produced a documentary of Enza Anderson's bid for the leadership of the Alliance party of Canada in the documentary, The Man in a Dress

more information
visit the site: James Wegg Review: Against All Odds
http://www.jamesweggreview.org/reviews/filmdvdvideo/a_man_in_a_dress.html

CHAPTER 11

The State of Sexual Reassignment Surgery in the Twenty First Century

For the vast majority of the pre-operative transsexuals who live in Toronto, Ontario, their transsexual journey does not necessarily end at the private Montreal estate of Drs. Claude Maynard and Pierre Brassard, however, it certainly passes through it. In the case of the male-to-female transsexuals, by the time they have arrived at the estate, they have chemically castrated themselves from the potent mix of synthetic estrogens and anti-androgens. In the case of the female-to-male transsexuals they have undergone the masculinizing process so far that they cannot present as normal females anymore. In the case of both male-to-female transsexuals and female-to-male transsexuals, by the time they have arrived at the estate, they are fervently committed to undergo their respective procedures. In the case of the male-to-female transsexuals whether a GRS procedure is in their health interests or not, they desperately want to undergo one. Whether a GRS procedure is in their health interests or not, they passionately rationalize how it is in their health interest to undergo one. In the vast majority of male-to-female transsexuals, they have their shtick down to an art. There is absolutely nothing the staff could say or document that they would require the candidate to sign —disclaimer or otherwise—that will dissuade them. Only a small percentage opt not to undergo the procedure at the last moment. Even in the vast majority of these exceptional cases, the ones who do opt out reschedule to try to go through with it again.

* * * * *

The earliest medical experiments, involving hormone replacement therapy began in Nazi German during the Second World War. Doctors in the Third Reich extracted male and female hormones from animals and injected them into the inmates of the death camps to study their effects. Much of the science that went on had little to do with the war effort and more to do with junk science. At the conclusion of the war the research and evidence into hormone therapy from the death camps was lost.

By 1950, the medical community was dispensing hormone extracts from animals and offering these medications to people for a variety of medical reasons. However, transsexual women obtained these medications (often illicitly), learning that the medications allowed them to grow breasts, soften their skin and over time develop the contours of born women. Also during the 1950's, modern surgical techniques made it possible to reconstruct the genital region of the human body (male and female). A few surgeons reportedly began exploratory surgeries to construct vaginas in male-to-female transsexuals by using skin grafts taken from the different regions of the body including thighs or buttocks, drawing upon then recent techniques for constructing vaginas in intersexed girls. [1]

Christine Jorgensen, a U.S. citizen, was among the first small group of male-to-female transsexuals to undergo this early version of what became known as "sex change" surgery. On December 1, 1952, the New York Daily News reported the fact that Jorgensen had undergone surgery to change sexes and Jorgensen, herself, became incorrectly known as the world's first transsexual to undergo modern "sex change" surgery, opening the doors of the imagination to male-to-female transsexuals everywhere. In any event the surgeons who performed the surgery were not accepting any more candidates at the time, making access to the surgery impossible.

In the surgery Christine Jorgensen underwent (which was rapidly evolving), doctors first removed the transsexual's male organs in one or more surgeries. The patient then waited through an extended period for healing. Then, in a surgery similar to those done to create vaginas for intersexed patients, surgeons constructed the patient's vagina by using skin grafts taken from the thighs or buttocks. In fact, in 1952, Christine Jorgensen has little more than what became known as a penectomy. Christine didn't obtain vaginoplasty surgery until 1954.

Although the patients reported being "pleased" with the result, the surgical technique had a number of problems. The skin grafts were reportedly unreliable and partially failed to "take". In effect the body rejected the new vagina. The skin grafts also left large scars in the region where the skin was removed. In addition, the patients of the first "sex change" surgery were left with little or no sensation in their surgically constructed vaginas.

In 1958, or thereabouts, Dr. Harry Benjamin, an endocrinologist, began to care for male-to-female transsexual patients, becoming the first American doctor to do so. Benjamin rejected the accepted thinking in the psychiatric community at the time that male-to-female transsexuals were by definition suffering from mental illness. Instead, he believed they were suffering from a cross-gender condition of unknown origins. He prescribed estrogens to a select group of his patients, who were pleading for medical feminization.

In 1958, or thereabouts, French plastic surgeon, George Burou, developed what is known today as the penile inversion technique—variations of which are still in wide use to this day. Dr. Burou's technique salvaged the male genitalia as a source of skin and sensitive erotic tissue to create the new female genitalia, including the vagina, the labia and the clitoris. At his clinic in Casablanca, Morocco, he performed his technique on hundreds of patients including a number of female impersonators. Coccinelle and April Ashley were well-known patients who went through Burou's clinic in Casablanca. A number of wealthy males (what the community would regard as the early transsexual admirers) sponsored the surgery from Burou for a number of female impersonators. They took on these male-to-female transsexuals as lovers. Aristotle Onassis was reportedly the most famous of the sponsors of "sex change" surgery. [1]

* * * * *

Religious groups, operating in the U.S., who learned of the Jorgensen story, pressed politicians to enact legislation to prevent plastic surgeons from practicing sexual reassignment surgery on U.S. soil. The U.S. College of Psychiatrists and Psychologists entered Transsexualism in the Diagnostics and Statistics Manual (DSM) as a disorder. The medical industry uses the DSM as a standard to form the diagnosis of patients. State institutions, with the aid of the families of transsexual family members, forced transsexuals (often children) into asylums to undergo treatment in the hopes of a cure. These asylums often administrated electroshock therapy to transsexual patients young and old with little effect. The people who felt gender dysphoria continued to feel dysphoria after all attempts at treating them were exhausted. [1]

* * * * *

There are no available statistics on the number of genital surgeries performed worldwide. However, one can estimate the total number of surgeries performed in the last twenty five years by taking the average number of surgeons in practice, multiplying that number by the average number of surgeries a surgeon performs per annum, multiplying that number by the number of years (25). Obviously less surgeons performed genital surgery in the sixties and seventies compared with the eighties and nineties.

Dr. Brassard performs approximately 200 GRS procedures per year. Let's use that figure as the average number of surgeries a surgeon performs per year.

At any one time there are approximately twenty three surgeons worldwide, who perform genital surgery, including Pierre Brassard, Toby Meltzer, Eugene Schrang, Marci Bowers, Gary Alter, Preecha, Chettawut, Pichet, Sanguan, Ruch Wongtrungkapon, James Bellringer, Timothy Terry, Phillip Thomas, James Dalrymple, Trevor Crofts, Michael Royle, Seghers, Peter Haertsch, Simon Ceber, Harold Reed, Michael Brownstein, Peter Walker, Jorge Sáenz. Meynard stopped performing GRS procedures shortly after the death of Michelle Renee, his patient.

However, there is a larger pool of surgeons today than in 1980. To compensate, let's say there is an average of twenty surgeons over the twenty five year time horizon.

Twenty surgeons, who on average perform two hundred surgeries per annum totals to four thousand per annum (4,000) multiplied by twenty five years totals to one hundred thousand surgeries. There are approximately 100,000 post operative transsexuals in the world, not counting deaths.

* * * * *

Dr. Brassard's technique is a variation of the technique Burou practiced in his medical office in Casablanca, Morocco as early as 1958. Dr. Brassard's technique is in contrast to Suporn's technique, which is largely based on skin grafts.

Dr. Brassard begins by opening the genital area in a "V" pattern to allow him to work on the genital fully exposed. Each opening along the stem of the "V" pattern is approximate five inches in length. Dr. Brassard then opens the scrotum. Inside the scrotum, Dr. Brassard, locates the cords to the testes, severs them and removes the testes, preserving and separating the nerves that travel through the scrotum. With the scrotum open, Dr. Brassard then detaches the skin of the penis, and spreads it out on a surgical table. He has an assistant remove the bulbs of the hair follicles on the skin of the penis from the inside out using a surgical device.

"The reason he removes the skin is to allow his assistant to remove the hair follicles while he continues working. The surgical device, itself, looks like a cross between a scraper and a shaver." says Michaela, a pre-operative transsexual woman, who witnessed Dr. Brassard perform the combination of a penectomy and vaginoplasty on her partner at the hospital in Montreal, Quebec.

172

After his assistant plucks the hair follicles from the skin of the penis (now separated from the body) Dr. Brassard returns to work on the shape of the skin of the penis, cutting away the head of the penis. Dr. Brassard then attaches the severed head of the penis to the body to form what will ultimately be the clitoris and labia. As Dr. Brassard sculpts the clitoris and labia, he bundles all the nerves that are now separated from the penis and scrotum beneath the newly created clitoris. Simultaneously, Dr. Brassard reduces the length of the urethra to a length appropriate for that of a neo vagina.

Dr. Brassard then returns to the skin of the penis, attaches it to the newly created labia, forming a cavity outside the body sculpted or shaped by a surgical packing that resembles a small condom.

"The surgical packing is white so that Brassard can see if there are any areas that are not stitched tight or closed." Michaela says.

As in any surgical procedure the patient bleeds profusely. Dr. Brassard counters any bleed using a laser to cauterize the sources of the bleeding.

"Brassard is very meticulous about that." Michaela says.

Dr. Brassard then takes the urethra and connects the end of the urethra to the inside of the exposed vaginal cavity. Dr. Brassard then stitches the skin of the penis around the surgical packing from the inside of what will be the vaginal cavity. Dr. Brassard stitches the skin onto the packing to ensure that it doesn't fall out post-operatively.

"This is the part that takes the most time." Kassandra says. Kassandra is a three-year post operative transsexual, who underwent what is known as gender reassignment surgery performed by Pierre Brassard, M.D. "Brassard can take hours just doing this one part of the procedure."

In Fact, Dr. Brassard performs over one hundred separate stitches in the course of the one GRS procedure.

When the neo vagina is completely formed, Dr. Brassard makes a vertical incision between two muscle groups where the scrotum and penis used to be attached to. Dr. Brassard, then, inserts the exposed vaginal cavity into the catheter by hand.

"He doubles checks it to make sure there is not an area that looks like it's good, but isn't good." Michaela says.

Dr. Brassard and his assistants require approximately forty five minutes to prepare the patient for what is known as a Gender Reassignment surgical procedure. Dr. Brassard requires no more than two hours to perform a GRS.

"Dr. Brassard was constantly asking me if I was Okay during the surgery and at a break and whether I was going to faint." Michaela says. "I was amazed at how meticulous Dr. Brassard is."

Transsexuals, who value sexual function and undergo a GRS procedure, are literally placing their faith in the hands of their surgeon. Prior to any surgical procedure Dr. Brassard's staff have every candidate sign legal documents, waiving Dr. Brassard from any and all liability. Although Dr. Brassard does not publish the success rates of his patients, who report being able to regain orgasmic function, Dr. Brassard's staff explicitly consult each candidate that there is a "very good chance" from undergoing GRS, that the candidate will fail to ever regain orgasmic function, assuming they were orgasmic prior.

"Dr. Menard's wife, who is one of the head nurses at the residence, in fact takes you aside and tells you there is a chance that you may never climax. They make it clear to that you may never function after the surgery. They tell you to [in effect] prepare yourself for the chance you may never sexually function moving forward." Kassandra says.

However, the sign of any good surgeon is whether people would otherwise engage his services. In the case of GRS, Dr. Brassard's skill is arguably in high demand. He sees over two hundred patients for GRS per year. There is little doubt that Dr. Brassard candidates come to him with their complete trust. They are steadfast undeterred by all warnings to the contrary that they may fail to ever regain orgasmic function after surgery, if that is indeed important to them. For most transsexuals, who undergo surgery, sexual function isn't important. In a recent published follow-up study (Radman, Lazer, Benet, Schaefer & Melman, 1999) involving a three year follow-up of 47 male to female post operative transsexuals, where only 28 patients were in fact contacted, of those, only 11 were contacted directly. Three are known to have died; one in a motor vehicle accident (the details of which are not disclosed), one from AIDS; and one from suicide in jail. Of the survivors, "all 28 expressed that they felt better from surgery." However, of those who said that orgasm was "very important for sexual satisfaction", four (29%) now report that they are unable to orgasm.

"He has a stack of endorsements from transsexual women, who write 'I want to thank you for 'saving' my life.'" Kassandra says.

"Would I have Dr. Brassard do my work? Absolutely." Michaela says, after witnessing his work firsthand.

However, Brassard's technique is like all penile inversion techniques in that it has limitations.

"He [Brassard] told me he would be able to give me a life-looking vaginal area. [including] a labia, a clitoris and a hood." Kassandra says.

After undergoing a GRS procedure, Dr. Brassard's patients will be at the hospital for approximately four days and nights. On the fifth day or thereabouts, Dr. Brassard's patients return to the residence. On the eighth day or thereabout, Dr. Brassard (or another staff member) removes the surgical packing.

When they are convalescing at the residence, the patients can make telephone calls and access the Internet and get back in the swing of things, figuratively. The nurses encourage Dr. Brassard's patients to take brief walks on the island of the residence.

"Even on the first walk, the nurses tell you not to go over the bridge." Michaela says. The residence is physically on a private island on the outskirts of Montreal, Quebec. "They tell you to stay on the island. Some of the girls walk vigorously [after the surgery]. We call it the 'Montreal shuffle'. We had a laugh. We said, "if it walks like a duck and it quacks like a duck, it could be a transsexual.' The one girl from Montreal, Anne, [who convalesced at the residence at the same time as Susan, Michaela's partner], was doing the 'duck'. She was over enunciating the walk and purposely being funny about it."

In the case of male-to-female transsexuals, who live in Ontario, Dr. Brassard is cost economical. Dr. Brassard is considerably less expensive than comparable U.S surgeons, including Dr. Eugene Schrang, Toby Meltzer and Marci Bowers, who now charges over US$15,000.00. Dr. Brassard only charges CDN$16,800.00 dollars for the GRS procedure, which includes post-surgical care for up to fourteen days at the residence. If one attends a consultation, that is deducted from the fourteen post-surgical days of care.

The expense to travel to Thailand to undergo GRS from Suporn is considerable. If you are transsexual and elect to undergo a GRS procedure with Suporn, you need to budget one month of post-surgical care, which is not included in Suporn's fees. The Hotel expense will run approximately US$1,000.00. The airfare to Thailand is approximately US$1,200.00. At the time of this writing Dr. Suporn was charging approximately US$12,000.00.

Dr. Brassard takes on a considerable number of patients, who are otherwise not satisfied with undergoing GRS from a third party surgeon. Dr. Brassard tends to these clients later in the work week, after he performs his scheduled load of GRS procedures.

* * * * *

Again, a gender identity disorder is marked by a persistent discomfort with one's gender role in society and a preference to live out another gender role. In theory the discomfort of having a gender identity disorder is largely ameliorated when one socially functions as a member of the opposite gender. In practice not so.

Males and females, who are young, can function as members of the opposite gender relatively quickly and with little medical intervention. Usually the introduction of hormones (hormone replacement therapy) is sufficient to allow a young male or female to socially appear and function as a member of the opposite gender. Male-to-female transsexuals and female-to-male transsexuals (even ones who are young) cannot actually arrive at being members of the opposite sex due to limitations in medical science. Transsexuals can sexually function as a member of the opposite sex in society only through surgical means. A gender identity condition in isolation is not powerful enough to motivate a transsexual to elect to undergo genital surgery.

The underlying motivations a sexually developed individual has towards undergoing genital surgery varies by sexual orientation. Heterosexual males, who have an irrepressible urge to undergo genital surgery, largely have a sexual arousal to being or turning into a woman. Homosexual males, who have an irrepressible urge to undergo genital surgery, largely have a sexual arousal to sexually functioning as a woman with another otherwise anatomical male. Often these homosexual males are not homosexual at all, but asexual, having little interest in third partners.

In the case of the homosexual, male-to-female transsexual, the urge or motivation in having a penectomy and vaginoplasty largely comes from the desire to sexually function as a member of the opposite sex. Having a surgically constructed vagina allows the androphilic male-to-female transsexual, to engage in intercourse with a member of the same anatomical sex (or in this case another male). The presence of a vagina alone is a shallow construct for socially functioning as a member of the opposite gender in society (or a woman in this case). However, few homosexual, male-to-female transsexuals ever explicitly state regret or revert to socially functioning as males in society. The fact that they do not is meaningless in the context of a behavioral addiction.

In case of the homosexual, female-to-male transsexual, the urge or motivation in having a surgically constructed penis is largely the same. The urge comes from the desire to sexually function as a member of the opposite sex (in this case male). Medical science offers little to satisfy this desire outside of cosmetic appearance. However, few gynephilic female-to-male transsexuals ever explicitly state regret or revert to socially functioning as females in society. The fact that they do not is again meaningless in the context of a behavioral addiction.

In the case of the heterosexual or bi-sexual, male-to-female transsexual, the urge or motivation in undergoing a penectomy and vaginoplasty largely comes from paraphilic tendency to become sexually aroused by the thought of being or functioning as a member of the opposite sex coupled with a behavioral addiction. Blanchard refers to this sexual phenomenon as autogynephilia and males that have this sexual predisposition as autogynephiles. Autogynephiles do not have a gender identity disorder. It only appears that way on the surface. A substantial number of gynephilic male-to-female transsexuals, explicitly state regret or revert to socially functioning as males (or at times butch dykes) or demonstrate complete gender apathy. A handful of masturbation thoughts during adolescence is all that is necessary to drive an autogynephile to the operating table. As the person with an affected autogynephilic disorder seizes onto the concept of a change in anatomical sex, autogynephilia takes the characteristics of a behavioral addiction. The behavioral addiction then drives the individual through undergoing a GRS procedure and not necessarily autogynephilic thoughts.

In the case of all types of transsexuals, the urge or motivation in obtaining sexual reassignment surgery can also occur in individuals who exhibit an obsessive behavior about becoming a member of the opposite sex (and are not satisfied with simply socially functioning as a member of the opposite gender) coupled with an affected autogynephilic disorder. These individuals suffer more from an obsessive compulsive disorder brought on by the behavioral addiction, underlying transsexualism, and not necessarily a gender identity disorder. People, who have an obsessive compulsive disorder, underlying transsexualism, have persistent, racing thoughts about undergoing a GRS procedure and often advertise that they will commit suicide if they are denied for the surgery. These people regard GRS procedures as necessary and often 'life-saving' surgery. In the case of the transsexual, who suffers from an obsessive compulsive disorder, a GRS procedure does little more than take away the object of the obsession. A GRS procedure may or may not help these people to get the monkey off their respective backs, however, these individuals often grab onto other obsessions down the road (like the ability to procreate as a female), because they never effectively deal with the underlying disorder. They treat the symptom and not the cause.

In the case of all transsexuals, the urge or motivation in obtaining sexual reassignment surgery can also occur in individuals who have a mood disorder, like depression coupled with an affected autogynephilic disorder. A common myth is that depression is embodied in a gender dysphoria. Dr. Anne Vitale's work (which contributes to this myth) is irresponsible in this regard. In fact, a gender dysphoria and depressive mood disorder are mutually exclusive prior to an invasive surgical procedure. Harbouring the belief that one can stave off depression by undergoing a GRS procedure is particularly self-destructive. In fact, transsexuals, who suffer from a depressive mood disorder, are no less likely staving off an episode of depression after undergoing a GRS procedure than not undergoing one at all. In fact the opposite is more likely. A penectomy is more likely to adversely affect the serotonin levels in the brain and cause depression.

"Anybody who changes their sex is doing it for sexual reasons," Nancy Nangeroni says. [2] Nangeroni, who is a male-to-female transsexual, was the host of the now defunct Gender Talk.

"There's just no way that you can change your sex for reasons that are not connected to sex. [A GRS procedure is] going to change your practice of sex, so there's no way that sex can be uninvolved in a change of sex. It's that big, invisible elephant standing in the corner of the room that nobody talks about, or very rarely. The notion that one can be transsexual and not do it out of sexual motivations I think is silly, because it's missing the obvious. And so I think Ray Blanchard thinking that he's saying something unique there is questionable." Nangeroni continues.

The reason why a large number of female-oriented, male-to-female transsexuals fail to benefit from genital surgery is that as the real (not explicit) drivers toward transition change so do the motivations to function as members of the opposite gender. The vast majority of males that have an affected autogynephilic disorder are utterly incapable of predicting the 'vanishing floor' that will down the road give way beneath their feet. In the case of the autogynephilic transsexual, as his sex drive wanes and the behavioral addiction subsides, he effectively reverts back to his fundamental gender identity, which is masculine. Autogynephiles, who identify as five years post-operative transsexuals, typically live as men or butch dykes or exhibit complete gender apathy. To say that there is no correlation between age, sex and sexual orientation of a transsexual and how well they progress into being well-adjusted member of the opposite sex in society is akin to saying there is no correlation between the sun and daylight.

Is there hope for the autogynephile? Absolutely. If the autogynephile can maintain a sex drive, if he can prop up his otherwise weak female gender identity with his natural endocrine system or supplement it with testosterone he can adequately function as a member of the opposite sex for a considerable stretch of time. However, inevitably the floor vanishes and even the once successful autogynephile realizes that his gender transition was a mistake.

Renee Richards, a female-oriented post-operative transsexual woman, is back and forth on whether or not her gender transition was a mistake. Richards on at least one occasion has stated on record that her gender transition was a mistake. Alternatively, the autogynephile can learn to be comfortable with her new gender role without the help of her sex drive. Living full time as a woman then becomes an awkward training exercise.

The motivations of a sexually developed individual who was misassigned at birth, like the case of David Reimer, is completely different. Reimer had a natural gender identity disorder. Reimer's motivations to undergo a phalloplasty procedure was to repair the damage of being misassigned at a young age. Reimer's disquiet about his assigned gender reportedly began years before his sexual development.

* * * * *

The likelihood of whether a transsexual will reportedly benefit from genital surgery is more a function of age, sex, and sexual orientation and less a function of the quality of the technique and/or the surgeon involved in the procedure. The regret rates in post-operative transsexual women have only marginally improved over the last forty years since the first skin grafts performed on transsexual women, dating back to 1962. The regret rates in post-operative female-to-male transsexuals are virtually non-existent. In the paper, Predicting Regrets in Post Operative Transsexuals, dated 1989, Blanchard, Dickey, Clemensen, Steiner report that out of seventy eight (78) male-to-female transsexuals surveyed, only 6% (4) either explicitly stated regret or reverted to living as men. Of the four individuals in the paper, dated 1989, who either explicitly stated regret or reverted to living as male, all had a heterosexual background prior to the surgery. Hence there is a significant correlation between one's sexual orientation and whether the person regrets the surgery. If the person regrets the surgery, it is reasonable to argue that the individual does not benefit from it as well. (See Appendix E— Maria: a Case Study in Post-Operative Regret).

Middle-aged, female-oriented transsexuals, who value sexual function, and are involved in committed relationships, should disabuse themselves of any pursuit of genital surgery. Few wives of middle-aged male-to-female transsexuals successfully cope with a husband's transition. In fact the vast majority leave their transsexual husbands, despite all attempts at relationship counseling.

"Can you blame them? They married a guy, who they thought were going to have babies with and help raise their kids with and grow old together. The male has now changed himself. If the wife is not at least bi-sexual, what have you become? Her friend?" Kassandra says.

In addition to the problems of retaining a relationship through the course of transition, a middle-aged, self-identifying lesbian transsexual woman, who values sexual function, has only a 71% likelihood of retaining sexual function after surgery. [3]

Self-identifying lesbian male-to-female transsexuals, who value sexual function, and who are otherwise open to a sexual relationship with a natal female (and not a male-to-female transsexual), should not pursue sexual reassignment surgery. Few homosexual women (particularly middle-aged, homosexual women) find male-to-female transsexuals attractive. In fact, less than 1% of all homosexual women prefer a sexual relationship with a male-to-female transsexual over one with a natal female.

"If you have had the surgery and you meet a girl and she goes down on you, she is going to say, 'you are not a woman'. Because the surgically constructed vagina is not [the same as] a genetic female's. It doesn't get as wet. There is no room for compromise on that." Kassandra says.

The self-identifying lesbian, post-operative transsexual, who is not attached, will simply fail to realize her expectations, involving relationships with natal women time and time again. Often, self-identifying lesbian, post-operative transsexuals (who do not otherwise exhibit gynandromorphophilia) compromise and take on other self-identifying lesbian, male-to-female transsexuals as lovers. Further, a significant number commit suicide, due to the onslaught of depressive episodes coupled with the realization of the true hardship at finding either a new full time position (if the individual is out-of-work) or a new intimacy partner or a combination of the two.

In theory homosexual male-to-female transsexuals are more likely to sexually benefit from genital surgery than their heterosexual counterparts. The surgery allows the homosexual, post-operative male-to-female transsexual to sexually function in an otherwise heterosexual relationship. In effect her neo vagina strengthens her chances at a relationship with a male who otherwise prefers vaginal intercourse with a transsexual. In theory, the genuine homosexual male-to-female transsexual's indigenous feminine gender identity is only enhanced by genital surgery and feels a greater sense of validation in society. In practice not so. Unfortunately, there are very few males who prefer sexual relations with a post-operative transsexual in lieu of a pre-operative, male-to-female transsexual or natal female. Further, few homosexual, male-to-female transsexuals value committed relationships and few engage in them for any significant length of time.

Even the homosexual, post-operative male-to-female transsexual, who finds 'herself' in a committed relationship with a so-called admirer, or a man who prefers sex with pre-operative male-to-female transsexuals over natal females, is more likely to survive in her relationships than her heterosexual, male-to-female counterparts, who are otherwise involved in a traditional marriage.

"If you are a [self-identifying lesbian] transsexual and you think you are going to get into a relationship with a genetic female or preserve your relationship with a genetic female, pass the pipe my way, [cause] I want try that drug. The reality is the percentage of women, who would continue in a relationship or be interested in a relationship is probably below one percent. (1%)" Kassandra says.

Autogynephilic male-to-female transsexuals, who have a paraphilic tendency to become aroused by the thought of being or turning into a member of the opposite sex, should disabuse themselves from pursuing genital surgery. Autogynephilic male-to-female transsexuals should seek alternative treatment for the behavioral addiction, underlying transsexualism, which is kicked off by a sexual response to the consideration of a change in sex (i.e., or similar irrepressible urge to undergo genital surgery). Few autogynephiles reportedly benefit from genital surgery and fewer still watch their respective lives improve. Further, few autogynephiles are able to maintain a sex drive to prop up an otherwise weak feminine gender identity.

The Transsexual Delusion

Male-to-female transsexuals, who have ancillary mental disorders, like obsessive compulsive disorder, bi-polar mood disorder or a personality disorder, should definitely not pursue genital surgery as no one knows how the various major mental disorders interoperate.

references

[1] LYNN CONWAY ARTICLE, VAGINOPLASTY: MALE-TO-FEMALE SEX REASSIGNMENT SURGERY

Lynn Conway publishes an article entitled, Vaginoplasty: Male to Female Sex Reassignment Surgery; Historical notes, descriptions, photos, references and links on her University of Michigan web site.
more information
Vaginoplasty: Male-to-Female Sex Reassignment Surgery
http://ai.eecs.umich.edu/people/conway/TS/SRS.html

[2] GENDER TALK BROADCAST #357

In her broadcast, dated April 29, 2002, Nancy Nangeroni elaborates on the relationship between sex and transsexual behavior. Nancy Nangeroni is quoted as saying, "[...] anybody who changes their sex is doing it for sexual reasons. There's just no way that you can change your sex for reasons that are not connected to sex." (17:25) Nangeroni continues, "the notion that one can be transsexual and not do it out of sexual motivations I think is silly because it's missing the obvious." (19:13) "the idea that auto -- excuse me, that gynephilic are doing it for more sexual reasons than [...] androphilic transsexuals is similarly silly." (19:21)

more information
transcript of gender talk broadcast #357
http://www.gendertalk.com/radio/programs/350/gt357.shtml

[3] Rehman J., follow up study on 28 post operative transsexuals

The reported sex and surgery satisfactions of 28 postoperative male-to-female transsexual patients; Archive OF Sexual Behavior; (1999) 28:71-89, Rehman J. Lazer S., Benet A.E., Schaeler L.C., Melman A.
more information
visit the site: Springlink Abstract of the Study
http://www.springerlink.com/content/p005571hmv827611/
visit the site: Springlink Orgasm in Post Operative Transsexuals
http://www.springerlink.com/content/p4347n067550n604/

CHAPTER 12

The Alternative Treatments to GRS Procedures

It is generally accepted in the psychiatric community that psychotherapy is an ineffective means of treating transsexualism. In his book, the Transsexual Phenomenon, Dr. Harry Benjamin emphatically states that psychotherapy in isolation is ineffective at treating transsexualism. Cognitive behavior therapy cannot in isolation sway an individual to accept their assigned gender role and/or sexual identity. The vast majority of psychiatrists believe the same can be said for a transvestia. What about chemical intervention? Is the pathology leading to transvestia or autogynephillic transsexualism—both of which are kicked off from a sexual response—treatable?

* * * * *

In 1987, Dr. Paul Fedoroff, who then worked at Johns Hopkins in Baltimore, Maryland , presented a case report of a 46-year-old man with a generalized anxiety disorder and transvestic fetishism who responded to treatment with buspirone. The case report involved Mr. A. (identity withheld), a 46-year-old, married heterosexual male with a generalized anxiety disorder. Mr. A. approached the John Hopkins Sexual Behaviors Consultation Unit at the insistence of his wife. He complained of difficulty becoming sexually aroused with his wife unless he was cross-dressed or fantasizing of being cross-dressed. At that time Mr. A. reported that he masturbated while cross-dressed four times per week. At the same time he reported having intercourse with his wife once every two to three months. Mr. A. denied that he had any difficulty with erectile dysfunction.

Mr. A. stated that he had always been an anxious person, even as a child. He felt anxious in situations from which escape would be difficult or in which he had little control of the situation. Specifically, he was unable to eat at public restaurants unless seated by a door. In addition, he was not able to take public transit. Ironically, he did not feel uncomfortable in social situations. He enjoyed being out dancing with his wife. He most frequently became anxious when he was by himself with nothing to do or when plans had changed. Although Mr. A. worked steadily, he complained of tension, autonomic hyperactivity, apprehensive expectations, and irritability. He said he dealt with this anxiety by drinking alcohol or by cross-dressing or both.

The clinicians at the Sexual Behaviors Consultation Unit performed an initial evaluation that included interviews with both Mr. A. and his wife. The clinicians diagnosed Mr. A. as having a transvestic fetishistic disorder, a generalized anxiety disorder, and an alcoholic dependence, according to the DSM-III-R criteria.

The clinician at the Sexual Behaviors Consultation Unit engaged Mr. A. and his wife in psychotherapy for two sessions. After that a single clinician engaged Mr. A. in psychotherapy on a weekly basis for the next two years.

The clinician initially focused on Mr. A.'s alcoholism, which the clinician concluded was a reflection of Mr. A.'s attempts at self-medication. As Mr. A. drank less alcohol, he reported more stress. The clinician started Mr. A. on alprazolam (0.25mg t.i.d.). Alprazolam is an anxiolytic medication which addresses anxiety disorders. This was increased to (0.25mg q.i.d.). At this level Mr. A. reported a decrease in stress. However, Mr. A.'s sexual proclivities towards cross-dressing remained unchanged. This observation was confirmed by Mr. A.'s wife, who began to attend the sessions with her husband as the psychotherapy shifted away from Mr. A's alcoholism to their personal relationship. Although Mr. A. reported that he now had no difficulty in restaurants or public transit, he continued to cross-dress when he felt tense. The clinician wished to increase the dosage of alprazolam. However, Mr. A. complained that the medication made him drowsy. He was also concerned about becoming chemically addicted to the medication.

The clinician then introduced buspirone at a dosage of 5 m.g. t.i.d. and increased the dosage to 5 m.g. q.i.d. The clinician weaned the patient off alprazolam within the first two weeks, after starting buspirone. Nine days after taking buspirone, Mr. A. reported a decrease in his urge to cross-dress. He completely stopped cross-dressing approximately three weeks later. Mr. A. and his wife reported no change in Mr. A.'s ability to become erect or his ability to reach orgasm and ejaculate while taking buspirone. The frequency in which Mr. A. and his wife had sex also increased from once every two to three months to once weekly. Mr. A. also claimed that he no longer masturbated. In his own words, Mr. A. reported that he was "astounded" that he "could become sexually aroused without cross-dressing or fantasizing about cross-dressing."

After three months of buspirone therapy and with the consent of Mr. A. the clinician took Mr. A. off busprione to see if his anxiety symptoms still required the medication. Within one week the patient reported feeling "edgier." He said that he had begun to have urges to cross-dress, and his wife reported a decrease in Mr. A.'s interest in having sexual relations with her, which Mr. A. confirmed. The clinician restarted Mr. A. on buspirone therapy with Mr. A.'s consent with results similar to the previous trial.

Dr. Fedoroff later contacted the patient in 1989 to learn how he was doing. Mr. A. said that he dropped out of the therapy at Johns Hopkins due to relocation and the inconvenience of attending the clinic. Mr. A. also said that he found alprazolam more effective than buspirone in reducing his anxiety but that only buspirone had any effect in reducing his urge to cross-dress.

Dr. Fedoroff noted that Mr. A. made two observations of note.

> He said that he continues to notice that he is more likely to cross-dress when he is anxious or "under stress." However, he has stopped taking any medication and has returned to his former frequency of cross-dressing because he finds it pleasurable and is now less concerned about the behavior since he feels he could reduce his urges by taking buspirone if he ever needed to.

* * * * *

Approximately 5% of all men in North America have a transvestia of one degree or another. Only in the rarest of cases does a man, who has an affected transvestic fetishistic disorder, regard the practice as an impediment to his day-to-day life. In fact the vast majority of males do not act out in public and the ones who do generally emerge as public cross-dressers late in life. Dr. Federoff describes this phenomenon as the 'aging' cross-dresser.

The 'aging' cross-dresser is one, who has accomplished all that he otherwise had hoped to accomplish, feels secure in his lot in life, is financially well off and otherwise has nothing to prove to anyone. In effect no one can touch the 'aging' cross-dresser or any of his "abhorrent" behavior as the result of public cross-dressing.

Only in the rarest of cases does a man, who has an affected transvestic fetishistic disorder, seek treatment and typically such a man only does so on the insistence of his common law spouse.

Today Dr. Federoff practices at the Sexual Behaviors Clinic at the Royal Ottawa Hospital as a forensic psychiatrist. There he continues to treat males, who seek to eradicate their urges to cross-dress. Dr. Federoff reportedly does not attempt to treat transsexuals with buspirone therapy.

* * * * *

In 1984 Dr. Elsie Shore published a case study on a 24-year-old male-to-female transsexual, who after living in the female role for two-and-a-half years reverted back to living as a male. Unlike a number of transsexual subjects, who reverted back to the male role after undergoing a gender reassignment surgical (GRS) procedure, Shore's subject, Michael a.k.a., Mickey, did not undergo a GRS procedure. At age 22 Mickey enrolled in a gender identity program for sex reassignment. Shore first met Mickey when Mickey agreed to participate in a research project. At that time Mickey had been undergoing hormone replacement therapy (HRT) for 21 months and had been living exclusively in the female gender role for 14 months. Mickey presented as a 'passable' female. Mickey was reportedly shy, lonely, and wanting to be loved and cared for.

Mickey was one of fraternal twins, the other a sister. Mickey also had a older brother. Mickey grew up in a Midwestern suburb. His father was a truck driver; his mother a housewife. Mickey reported that he often slept in the same bed as his older brother. Mickey said he wanted to marry a man, who is otherwise kind and non-domineering like his brother.

Mickey reported wanting to be like other girls, starting in the second grade. He began to cross-dress at that time, starting with the clothes of his twin sister. Mickey's proclivities towards cross-dressing waned in Junior High School, due reportedly to privacy issues. Mickey was active in sports and elevated to being captain of the football team as well as class president. Although the girls in his class found him attractive, he never took on a girlfriend.

Although Mickey cited an interest towards transsexualism as the reason for turning away female advances, he reported being shy his entire adolescence. Mickey first identified as a transsexual after watching a program on the subject. After watching the program Mickey reportedly masturbated for the first time. His masturbation thoughts involved changing genders and living as a female. Mickey's entire sexual history involves only two incidents; one with a male; and one with a female. In each case the other party made the advances.

After High School Mickey went to College and earned a College degree.

Mickey reportedly suffered from depression while in school after learning he had been called up for military service. Mickey's depressive episode reportedly lasted two years. Mickey did not seek any medical intervention for his depression. The military gave Mickey a 4F, based on his transvestism—that meant he didn't have to serve.

After graduating from College Mickey immediately applied for a sex reassignment. The gender identity clinic thought Mickey was a good candidate to go through the gender transition and authorized him to undergo HRT. The gender identity clinic that Mickey went to comprised a psychologist, psychiatrist, obstetrician-gynecologist, and endocrinology and surgery consultants. The team followed the harry Benjamin Foundation's Standards of Care. Specifically, the psychological test battery that Mickey went through had included the MMPI, 16 PF, Rotter Sentence Completion Test, House-Tree_person Technique, and the Gender Identity Checklist.

To qualify for a GRS procedure, which Mickey sought, Mickey had to live in the female role for one to two years. After acceptance into the program Mickey began to undergo HRT. He also protracted his facial hair through electrolysis. Mickey had little trouble adjusting to his new role as 'Mickey'. He found employment at a dairy store and made friends with other women and with men.

Finally, the day before Mickey was to undergo the GRS procedure, he was notified that the governing body of the hospital had decided to discontinue such procedures. Although he felt frustrated by the news, he was somewhat relieved. When asked what he thought would have happened if he had had the surgery, Mickey said that he would have continued living as a woman. However, Mickey also reported that he did not think that he would have been as happy as he now feels he can be as a man.

Mickey attributes part of his decision to revert to his original gender role to his vocational plans. Training in nursing had been a goal for some time, and he was scheduled to begin his studies after the surgery. The school that had accepted him as Mickey—the female—did not know of his transsexuality. Earlier, a different nursing school had made completion of surgical reassignment a prerequisite for acceptance. Mickey reported that he did not feel the drive to undergo the GRS procedure as strongly after the news of the unconditional acceptance by the second school. Shore noted the flaw in this reasoning on Mickey's part as the second school was in fact ignorant of Mickey's physical state.

Shore noted that the more realistic explanation of Mickey's change in attitude regarding the GRS procedure is that Mickey had had doubts about the surgery before it was canceled. Mickey's attitude towards the surgery, however, was at the least laissez faire.

Outside of Mickey's vocational ambitions Mickey, while still living in the female role, joined a charismatic religious movement. In addition to the resurgence of religion in his life, Mickey reported that he found a different kind of man among the church members. Mickey saw the men in the church more like him—gentle and passive without the loss of masculinity. Mickey also became involved with a woman at work. Given his so-called Gender Identity Disorder, Mickey did not view his attractions to the woman as homosexual but rather as the need to return to the male role. Approximately two months after Mickey's surgery had been cancelled, Mickey reportedly began to discuss the possibility that he may revert back to living as a man to the woman, he felt enamored with. Mickey stopped taking female hormones at that time, although he had not been taking the medication regularly for a while.

* * * * *

Michael, a.k.a., Mickey, reported to Dr. Shore that he began to cross-dress in second grade at age 8 or thereabouts. The sexual identity of an 8-year-old child is fully developed and immutable. Michael cannot have a gender identity disorder and any diagnosis to that effect is fundamentally flawed.

Michael reported notable High School accomplishments as a male, including being selected as the captain of the football team. Playing football is a masculine interest. Michael also reported a indifferent sexual history. Michael's interest in men as marriage partners had little to do with any one particular man and more to do with functioning as a female in a working relationship, which is asexual and not homosexual. Michael reported that after watching a show about transsexualism, he masturbated to the thought of affecting a gender transition. Masturbating to the thought of affecting a gender transition is autogynephilic in nature and not indigenous to females. Michael perfectly fits the profile of an asexual autogynephile.

Michael suffered from a considerably long episode of depression which began in his final year of College. Michael reportedly took nothing for the depression. In the absence of an anti-depressant medication or Selective Seratonin Reuptake Inhibitors (SSRIs), the body relies on its own natural mechanisms to bring itself back into balance.

50% of all the people, who suffer from clinical depression, produce excess levels of cortisol, a hormone found in the blood system. In fact, the hypothalamus kicks off the process by producing corticotrophic-releasing hormone (CRH), which in turn stimulates the adrenal glands, located near the kidneys, to produce the cortisol and release the chemical into the blood system. The pituitary gland in turn produces adrenocorticotrophic hormone (ACTH). ACTH has anticonvulsant effects on the human brain, forcing the nucleus accumbens of the brain back into balance. [1] The nucleus accumbens of the brain is the heart of the brain's reward system. That is one reason depressive episodes have cycles with definitive beginnings and ends. As Michael's own body produced ACTH—a natural anticonvulsant—to counteract the symptoms of depression, the ACTH had the tertiary effect of ameliorating a faulty reward system in his brain, bringing it back into balance.

By the time Michael's scheduled surgery date arrived, his interests in a gender transition had effectively evaporated whereas most candidates for GRS procedures, who otherwise do not suffer from a protracted episode of depression, would be livid. On the event of meeting a woman he felt enamored with he stopped the medication protocol and reverted to living as Michael.

On reading Shore's case study of the former transsexual, a number of the centralized gender identity clinics, including the Centre for Addiction and Mental Health (CAMH), began to dispense SSRIs as an alternative treatment to GRS procedures. However, they were off the mark as SSRIs are largely ineffective at treating behavioral addictions, including the behavioral addiction, underlying transsexualism. Ultimately, the centralized gender clinics returned to aversion techniques and nothing of value came out of Shore's case study.

* * * * *

To find effective treatments for mental disorders, you must often go beyond what is publicly available and immerse yourself amongst the inflicted. If you want to find people, who by accident undergo effective treatment for a mental disorder, the best places to start are the support centres. In the case of transsexualism, the best places to start are the transition support groups. These volunteer-run, support groups see people with various ancillary disorders in addition to what is on the surface. These groups see people in all different phase of transition and with various degrees of the disorder. It is no surprise, then, that they see and hear of people, who suddenly and silently lose interest in gender transition. Behind every individual, who loses interest and drops out, there is a story and that's where you learn not only the nature of the disorder, but what treats it. When you discover a case example, you must then cross-correlate that to the research by clinical researchers.

Clinical researchers are medicine's detectives, cross-correlating mental disorders in studies in an attempt to isolate effective treatments for various disorders. Clinical researchers and not medical doctors discovered that Epileptics, who are successfully treated for epilepsy, have a lower incidence of suffering from manic episodes. As a result doctors prescribe divalproex sodium to patients, who suffer from affected bi-polar disorder. Divalproex sodium was otherwise developed to ameliorate epileptic seizures. However, clinical researchers need people in the field to test their hypothesis. That's where transsexuals come into play.

Clinical researchers combined with transsexuals and not medical doctors, are indirectly responsible for the subject of a case study, who silently and without fanfare was effectively treated for the behavioral addiction, underlying transsexualism, and dropped out of the community of transsexuals with his genitalia intact.

From 1998 through 2005, Dr. Frank Emil Cashman, a psychiatrist at the St. Michael hospital, treated a patient with a "mild" mood disorder and a gender identity disorder, who in the case of the gender identity disorder incidentally responded to treatment with divalproex sodium. The patient, who was very active in the transsexual community and attended support group meetings, suddenly and without explanation dropped off the radar screen. The case study is not reported in any medical or scientific journal, yet the case is significant. In fact, Dr. Cashman never sought accolades for his success on the case matter and none has ever been awarded to him. The case study is otherwise completely undocumented.

In May, 1998, Dr. Cashman took on G. (identity withheld), a 38-year-old, divorced, heterosexual male, who had by that time been diagnosed by a forensic psychiatrist as having a bi-polar mood disorder and a transvestic fetishistic disorder. When Dr. Cashman took on G.'s file, G. had by that time reportedly suffered from a single manic episode. In 1996, G. had gotten into trouble with the law on at least one occasion and following an arrest in 1996, had found himself incarcerated. In fact the episode was so intense that G. had to be hospitalized in an institutional prison over the course of a number of weeks and had to be medicated with

Lithium. Prior to the arrest in 1996, G. emerged as a public cross-dresser and experimented with estrogens to change his physical appearance.

Later, a second forensic psychiatrist who looked at the case file changed the diagnosis to that of a drug-induced, bi-polar mood disorder secondary to Zoloft, and a transvestia. Later, a third forensic psychiatrist involved in the case file, changed the diagnosis again to that of a drug-induced, bi-polar disorder, secondary to estrogen and a gender identity disorder. Prior to the arrest, G. had been seeing a board certified psychiatrist to treat what presented as a 'mild' depression with Zoloft—an anti-depressant medication. It is now known that Zoloft, which G. took in 1996, causes the adverse affects of mania although that was not known when the psychiatrist prescribed the medication to him.

G.'s mother, W., reported to the second forensic psychiatrist that G. otherwise grew up in a privileged environment. W. further reported that G.'s childhood was relatively normal, save one incident where G. had fallen off a high chair as an infant and suffered considerable trauma to the head. W. further reported that G., her son, never got into any trouble with the law and was otherwise a 'good' boy. W. further reported that G. was an introvert and, like his father, kept to himself.

Dr. Cashman took G. off Lithium, which he felt was not necessary for a person, who had only suffered from a single manic episode coupled with the evidence that it appeared to be triggered by a prescription medication.

However, Dr. Cashman had to be very mindful of G.'s family background to hedge off any reoccurrence of mania. G. reportedly had a family history of mood disorders. G.'s maternal grandmother had two episodes of depression late in life. G.'s mother reportedly suffered from a chronic anxiety disorder most of her adult life. Nothing was known of G.'s father at the time Cashman took over G.'s file as G.'s father and G. were estranged from each other, when G. was an adolescent. G. only discovered years later in September, 2005, on the news of his father's death, that G.'s father, J., suffered from an alcohol addiction and that he was in fact a life-long alcoholic. Given G.'s behavioral addiction, underlying transsexualism, which emerged late in life, J.'s alcoholism in retrospect was a dramatic discovery. Further, G. emerged from the arrest in 1996, without a criminal conviction and Dr. Cashman didn't want to see G. get into further trouble with the law. Given these considerations, Cashman dispensed divalproex sodium to G. on an occasional basis and only when he observed G.'s mood elevate.

In learning of G.'s proclivities towards cross-dressing, Dr. Cashman asked G. to report whether he publicly cross-dressed and to what degree and if he was undergoing hormone replacement therapy, which he had undergone briefly in 1996. Dr. Cashman also asked G. to disclose any drugs he was taking.

In fact G. did not have a substance abuse problem and when starting in 1982, G. cut out all illegal drug use.

Dr. Cashman was very perplexed by G.'s apparent gender identity disorder as G. showed no outward signs of femininity, save his eyebrows, which he plucked. In fact G. was straight and was by the time Dr. Cashman saw him involved with H., a woman four-and-a-half years his senior. Further, G. reported that he cross-dressed in early childhood and again in the early phases of adolescence, but G. only rarely cross-dressed as an adult and only during traditional events that see men cross-dress, including Halloween.

Dr. Cashman didn't probe G. on his gender identity disorder in any great detail as he recognized that he was not a subject matter expert on transsexualism and routinely referred patients, who exhibited transsexual proclivities, to Dr. Ray Blanchard, Robert Dickey, M.D. and the gender identity clinic at the CAMH—a first wave provider of transsexual care.

In 1997, G. re-emerged as a public cross-dresser after his troubles with the law reached a final conclusion.

In 1997, G. had applied for an assessment at the gender identity clinic at the CAMH, but never followed up with it. Instead, G. attended the offices of a number of private doctors and endocrinologists, who prescribed G. feminizing medications.

In 1997, G. saw Dr. Heather Davies, M.D., a general practitioner, who operated a gender clinic. G. learned about Davies' clinic through Patricia Aldridge. Under Dr. Davies' care G. underwent hormone replacement therapy. In 1998, G. unilaterally stopped HRT, suspecting that he did not have a gender identity disorder at all, but rather a sexual fetish to the thought of being female.

In 1999, G. went back to Dr. Davies with renewed interest in transsexualism, and went back on HRT. This time G. stayed on HRT under Davies's watch until December, 1999, when Davies closed her gender clinic. At the time Davies closed her clinic, Davies referred G. to endocrinologist Betty Chan, who G. liked, and G. picked up his medication protocol with Chan where he left off with Davies. In addition to undergoing HRT, G. underwent laser treatment to protract his facial hair. Although he successfully removed a portion of his facial hair from the laser therapy, he plucked the remaining follicles from his face on a weekly basis in an effort to look more passable.

In February, 2000 G., who responded to the HRT, separated with his girlfriend, who he had otherwise lived with for three years and began to live as his feminine alter ego on an experimental basis. The separation was bitter, yet amicable. Although H. attempted many times to dissuade G. from taking steps towards a gender transition, she came to the realization that G.'s proclivities towards transsexualism were irrepressible. Although G. mourned the demise of the relationship with H., G. was very excited with his decision to transition and felt it was the correct one.

G. as his feminine alter ego was not a particularly attractive male-to-female transsexual, but nevertheless G. as his feminine alter ego was visibly transsexual and reportedly people had no problem addressing G. with the appropriate gender pronouns wherever he went.

In May, 2000, after briefly living full time as his feminine alter ego for three months, G. reportedly fell ill with a mild depression and stopped undergoing HRT. G. blamed the depression on the synthetic estrogens. In November, 2001, however, G., who fully recovered from what turned out to be a brief episode of depression started back on HRT, only to stop again in 2002 due to indifference.

Sometime later in 2003, G. approached Betty Chan to restart the HRT, believing that he had a grasp on his true feelings, which were to go through with the gender transition, but Chan refused. Instead, Chan referred G. to the CAMH.

Hearing of the horror stories from other transsexuals reported, G. refused to go to the CAMH and instead went to the Sherbourne Health Centre, where he began to see one of the clinicians there. In January, 2004, the clinician at the Sherbourne Health Centre put G. back onto HRT on his persistent wish to be female and transsexual. Simultaneously, G. announced to his former wife and child that he would affect a gender transition and re-emerge in society as his feminine alter ego, taking only a brief holiday from work and his financial support obligations to his family.

Just as G. appeared to be back on the gender transition track and happy again in June, 2004, G. again reportedly fell ill to a mild depression and elected to discontinue the hormone replacement therapy.

In June, 2004, or thereabouts, Dr. Cashman, who took detailed notes of G.'s gender transition, observed that G.'s flip-flopping interest in HRT was strongly correlated to the introduction of divalproex sodium that Cashman prescribed to treat G.'s mood fluctuations. Again, Cashman had had G. on divalproex sodium, a medication, ordinarily prescribed to treat epilepsy, but in G.'s case Cashman had G. on divalproex sodium to treat G.'s mood fluctuations. As Cashman introduced divalproex sodium G.'s interest in transsexualism waned. Once Cashman saw G.'s mood level out, Cashman took G. off the divalproex sodium. As Cashman stopped prescribing G. divalproex sodium G.'s interests in transsexualism re-emerged months later.

G.'s general mood reportedly improved over time and once G. was clear of divalproex sodium for more than one year, G. reported that he no longer had interests in undergoing hormone replacement therapy or being transsexual. Today, G.'s transsexual experiences as his feminine ego are a faint memory.

*　*　*　*　*

Medical researchers have shown that hereditary plays a significant role in addiction. The gene is passed from father to child. The propensity for the gene associated with addiction to pass to male offspring is higher than female offspring. It is unknown how big the extra propensity is, but the children of parents with alcohol addictions, for example, must keep in mind that they have a greater chance to develop problems with alcohol and other behavioral addictions. Some studies suggest that the risk is 50% over the children of parents, who have no alcohol addictions. It is also known that certain personality characteristics, which are partly inherited, influence the risk of getting addicted. People who are often anxious who seek excitement in life and who are more antisocial are more likely to have alcohol addictions.

G.'s father J. was reportedly a life–long alcoholic. G. reportedly grew up an introvert. Given G.'s father had an alcohol addiction, given G. was reportedly an introvert during adolescence, G. has a much higher propensity to succumb to some form of addiction than the norm.

The higher propensity of the gene to pass from father to son rather than from father to daughter could explain why transsexualism as a behavioral addiction has a far greater incidence in males than females. The ratio is also significant. The male-to-female ratio of pathological sex addiction is 3 to 1 (3:1)—virtually the same male-to-female ratio of transsexualism. Further, the male-to-female ratio of pathological gambling addiction is also approximately 3 to 1 (2.8:1)—again, virtually the same male-to-female ratio of transsexualism. [2] The male-to-female ratio of pathological transsexualism is 3:1. [3]

* * * * *

The body of evidence that links epilepsy (and the effective treatment thereof) and sexual paraphilias (including transsexualism) is overwhelming. Hoenig and Kenna [1979] reported a strong correlation between lesions on the temporal lobe and sexual deviancy (including transsexualism). [4] Hoenig et al. cite Külver and Bucy's observations of hypersexuality in monkeys following bilateral temporal lobectomy. Terzian [1958] reported similar observations in humans. However, in a study of 86 men with predominantly temporal lobe epilepsy, Kolársky et al. [1967] found no evidence of hypersexuality. In fact Kolársky et al. [1967] examined the relationship between epilepsy and sexual paraphilias more systematically. Out of a sample of 86 men, four-out-of-five showed no signs of sexual abnormalities, but the researchers found that sexual abnormalities were more likely to be present if the cerebral lesion was present within the first year of life. The implication is simple. If the child suffers a trauma to the head as an infant, the child is more likely to develop paraphilic interests, including cross-gender proclivities.

G.'s mother, W., reported that G. fell off a high chair at a young age and that the incident was traumatic. If so, G. may have a legion from the event, setting the stage for sexual paraphilic tendencies.

There is a significant and material absence of transsexualism in epileptics, who are successfully treated for epileptic seizures. In a paper in 1963, Hunter et al. reported on a transsexual subject who developed seizures in his temporal lobe region of the brain at age 29. When the physician administered anticonvulsants, both the subject's seizures and his proclivities towards transsexualism reportedly waned. [5] The subject proclivities towards transsexualism reportedly developed at age 9.

In 1965, clinical researcher Jan Wälinder reported on a 29-year-old male, who had a normal development up to age 23, when he began a gender transition. Wälinder diagnosed the subject to have a predominantly left-sided EEG abnormality, consisting of bursts of theta and sharp waves. When treated with anticonvulsants his proclivities towards transsexualism waned, only to return when the medications were stopped. [6]

In his research paper, Transsexualism, a Study of 43 Cases, dated 1967, Jan Wålinder, reports that out of 207 subjects who exhibit cross-gender proclivities 7 report having epilepsy. The World Health Organization (WHO) reports that epilepsy occurs in 8.2 persons in 1,000 or 0.82%. The number is higher in developing countries. The incidence of epilepsy in population of 207 subjects who exhibit cross-gender proclivities is 3.38%—far greater. However, Wålinder reported that in all 3 case studies, involving epileptic subjects, none reported any signs of epilepsy, which suggests that although a small number of subjects had epilepsy, none was being treated for epileptic seizures. [7]

G. was never diagnosed as having epilepsy, but Dr. Cashman did administer divalproex sodium—an anticonvulsant—to G. to attenuate G.'s apparent mood fluctuations. Each time Cashman introduced divalproex sodium to G., G.'s interest in transsexualism waned, whether he suffered mild depression or not. As Cashman took G. off the medication, G.'s interest in transsexualism re-emerged.

Blumer [1969] reported two of fifteen transsexuals, who had abnormal EEGs. However, Blumer discussed three patients, who had both epilepsy and transsexualism and concluded that epilepsy is uncommon in transsexuals, but could occasionally be closely associated with it. [8] In his paper, dated 1969, Blumer called for more research into the relationship between epilepsy and transsexualism. None was ever embarked on.

In treating G.'s mood fluctuations Cashman accidentally treated G. for the behavioral addiction, underlying transsexualism.

Dr. Cashman's experience with G. is not unique, but is typical of individuals, who take anticonvulsants to treat an ancillary disorder. As the incidence of transsexualism—which is a behavioral addiction kicked off by a sexual response—is absent in people, who are successfully treated for epileptic seizures psychiatrists are unknowingly treating a host of underlying disorders in their patients, possibly including sex addiction, Internet addiction, gambling addiction and any disorder that is manifest by a faulty reward system in the human brain.

references

[1] BURNHAM, LONSDALE, SHAHZAMANI, PEREZ-CRUZ, EDWARDS: DEVELOPMENT OF NEW ANTICONVULSANTS USING THE KINDLING MODEL

BURNHAM, LONSDALE, SHAHZAMANI, PEREZ-CRUZ, EDWARDS: (2005) Development of New Anticonvulsants Using the Kindling Model; Advances in Behavioral Biology, Volume 55, 325-332.

[2] WENZEL, HANNE GRO, DAHL, ALV A.: FEMALE PATHOLOGICAL GAMBLERS—A CRITICAL REVIEW OF THE CLINIC FINDINGS

WENZEL, HANNE GRO, DAHL, ALV A.: (2008) Female Pathological Gamblers—A Critical Review of the Clinic Findings. International Journal of Mental Health & Addiction, January 12, 2008, ISSN: 1557-1874 (Print) 1557-1882 (Online); DOI: 10.1007/s11469-008-9174-0.

[3] EKLUND, P.L, GOOREN, L.J., BEZEMER, P.D.: PREVALENCE OF TRANSSEXUALISM IN THE NETHERLANDS

EKLUND, P.L, GOOREN, L.J., BEZEMER, P.D.: (1988) Prevalence of Transsexualism in the Netherlands. British Journal of Psychiatry, 152: 638-640 (1988)

[4] HOENIG, KENNA: EEG ABNORMALITIES AND TRANSSEXUALISM

HUNTER, R., LOGUE, V. & MCMENEMY, W. H. (1963) Temporal lobe epilepsy supervening on longstanding transvestism and fetishism. Epilepsia, 4, 60-65

[5] HUNTER, LOGUE, MCMENEMY TEMPORAL LOBE EPILEPSY

HOENIG, J., KENNA, J.C.: (1979) EEG Abnormalities and Transsexualism. British Journal of Psychiatry, 134, 293-300.

[6] JAN WÅLINDER TRAVESTISM, DEFINITION AND EVIDENCE IN FAVOUR OF OCCASIONAL DERIVATION FROM CEREBRAL DYSFUNCTION

JAN WÄLINDER: Transvestism, Definition and Evidence in Favour of Occasional Derivation from Cerebral Dysfunction. International Journal of Neuropsychiatry, 1,567–573.

[7] JAN WÅLINDER TRANSSEXUALISM: A STUDY OF 43 CASES

In his study of 43 Cases Jan Wålinder reports a low incidence of epilepsy in transsexual patients. Wålinder reports 7 cases of transsexual patients with epilepsy, but also reports that their cases did not require seizure medication.
more information
visit the site: A Study of 43 Cases
http://www.symposion.com/ijt/walinder/

[8] BLUMER: TRANSSEXUALISM, SEXUAL DYSFUNCTION AND TEMPORAL LOBE DISORDER

BLUMER, D. (1969) Transsexualism, sexual dysfunction and temporal lobe disorder. In Transsexualism and Sex Reassignment. Ed. R. Green and J. Money. Baltimore: Johns Hopkins Press, 213-219.

CHAPTER 13

Closing

Self Identity is a powerful concept in the advancement of human liberties and freedoms. However, self identity in the context of science and medicine is founded on self-diagnosis. Self diagnosis, by the otherwise layman individual is no substitute for proper diagnosis. The acceptance of collective self diagnosis runs counter to the foundation of scientific research and understanding. Whereas the medical community does not and should not listen to the anorexic community in regards to the approach of their health care needs, the medical community should not acquiesce to the political demands from LGBT groups that put self identity and market demand ahead of the Hippocratic Oath.

Sex is a powerful motivating force. "Sex puts the groove in one's 'tude." says Lace Dhalila Amour. [1] Amour is a public cross-dresser from Rhode Island, who travels to Toronto, Ontario for non-business related reasons frequently.

Sex and the anticipation of sex can cause autonomic and hormonal changes in people's bodies. Sex and the anticipation of sex can cause a woman to dress up for an evening with a lover. Sex and the anticipation of sex can cause a woman with a masochist disorder to willfully submit to being chained, humiliated, flogged and beaten. Sexual arousal ultimately produces endorphins and ultimately euphoria. Sex and the anticipation of sex can infuse obsessive compulsive thoughts, which can linger long after the point where the individual no longer derives any euphoria from the sexual stimulus.

Even leading transsexual advocates acknowledge the connection between sex and the motivations underlying a gender transition.

"Anybody who changes their sex is doing it for sexual reasons," Nancy Nangeroni says. [2] Nangeroni, who is a male-to-female transsexual, was the host of the now defunct Gender Talk.

Sex is an irrepressible periodic urge. In the context of a change to a non-intersexual's anatomy sex (definition 3) cannot be uninvolved. The fact that sex cannot be uninvolved is the fundamental flaw to the supposition that a disconnect between gender identity and sexual reality is the cause of transsexual behavior. Further, the primary sex organs of a person do not weigh into social functioning and hence gender identity in isolation cannot motivate anyone to undergo a GRS procedure. The only phenomenon remaining that can motivate a person to undergo a gender reassignment surgical (GRS) procedure is the obsessive compulsive thoughts, the behavioral addition, brought on by the consideration. David Reimer's case is different. Reimer, who was unaware of his true biology until age 11, sought to extricate himself from the constant teasing by undergoing a phalloplasty procedure—the only option available to him.

Despite the overwhelming quantitative evidence that transsexuals, who undergoing GRS procedures, 'feel better' post-operatively, their lives and in particular their health invariably get worse and not better. The ultimate measure into the effectiveness of any treatment is whether the individual performs better having undergone the treatment as opposed to not. The ultimate measure of the effectiveness of any treatment is not whether the person simply reports that they 'feel better' after undergoing the treatment, particularly if there is a complete disconnect between perception and reality.

Given sexually developed people, who undergo GRS procedures, watch their lives get worse and not better, GRS procedures are not an effective treatment for any inhibiting disorder, Gender Identity Disorder (GID) or otherwise. In fact GID in sexually developed individuals is fundamentally flawed. GID as it is articulated in the DSM IV is inadequate to explain the landscape of transsexual narratives.

Blanchard's concept of autogynephilia more closely fits the vast majority of heterosexual transsexual life narratives and not gender identity disorder. Further, the disorder also more closely fits the male-to-female transsexual narratives of men, who appear homosexual on the surface, but are in fact asexual in their sexual practices—deriving sexual arousal from the thought of functioning as a female in the early stages of the disorder regardless of the physical sex of a partner. However, Blanchard's concept of autogynephilia has limitations. In the latter stages of the disorder a behavioral addiction and not autogynephilia takes over and ultimately drives the person through the various phases of the disorder.

Auto eroticism, e.g., autogynephilia in males, autoandrophilia in females coupled with behavioral addiction completely explain all male and female transsexual life narratives and not gender identity disorder.

Similarly, the concept of autoandrophilia is more consistent with the significant number of heterosexual, female-to-male transsexual narratives and again not gender identity disorder. Further, autoandrophilia more closely fits the transsexual narratives of women, who appear homosexual on the surface, but are in fact asexual in their sexual practices. The discontinuity between the real life narratives of female-to-male transsexual and the accepted understanding of transsexual behavior in females warrants more formal scientific research. In any event in the case of latter stages of the disorder a behavioral addiction takes over from autoandrophilia and drives the inflicted female through the various invasive surgeries.

Addiction, including behavioral addiction, is hereditary. The gene that affects addiction has a higher propensity to pass from father to son rather than from father to daughter. The tendency for the gene to pass from father to son rather than from father to daughter could explain why transsexualism as a behavioral addiction has a far greater incidence in males than females. The ratio is also significant. The male-to-female ratio of pathological sex addiction is 3 to 1 (3:1)—virtually the same male-to-female ratio of transsexualism. Further, the male-to-female ratio of pathological gambling addiction is also approximately 3 to 1 (2.8:1)—again, virtually the same male-to-female ratio of transsexualism. [3] In any event more research is needed in this area to determine the cause of the disparity in the incidence of transsexualism between men and women.

Gender Identity Disorder occurs naturally in intersexed individuals, who are assigned at birth (and in the exception circumstance, like that of David Reimer). In this small number of cases, a GRS procedure is warranted and appropriate. However, in the case of the vast number of males and females, who seek GRS procedures, the procedure is not warranted, nor appropriate as it creates more health problems than it addresses superficially. In the case of the heterosexual and asexual autogynephiles, a GRS procedure is definitely not warranted, nor appropriate as it corroborates with the behavioral addiction rather than treating it. In the case of the heterosexual and asexual autoandrophiles (female-to-male transsexuals), more research is needed, but in the interim medical practitioners should err on the side of caution and not approve this group to undergo phalloplasty procedures.

Autogynephilia and autoandrophilia are one distinct disorder independent of sex—autoeroticism, e.g., automorphophilia. Autoeroticism, e.g., automorphophilia is the sexual response in men and women to the consideration of being or turning into a member of the opposite sex.

Autoeroticism, e.g., automorphophilia, which is sexual in nature, takes the shape of a behavioral addiction early in its lifecycle, like that of a gambling addiction, Internet addiction, shopping addiction, or sex addiction.

In any event, an affected autoerotic addiction (autogynephilia in males and autoandrophilia in females) is easily treatable. The drugs and medications that treat epileptic seizures and chemical and behavioral addictions, including methamphetamine addictions, cocaine addictions, caffeine addictions, sex addictions, gambling addictions, etc., effectively treat autogynephilia in men and autoandrophilia in women, including Vigabatrin. Vigabatrin, an antiepilepsy drug that boosts GABA (gamma-aminobutyric acid) in the synapse, is available in 60 countries worldwide, but not in the U.S.

Health jurisdictions should divorce themselves from the Harry Benjamin International Gender Dysphoria Association (HBIGDA) and affect sweeping and dramatic changes to the Standards of Care, requiring transsexual candidates who seek a GRS procedure to undergo non-invasive therapy, involving medications that directly deal with behavioral addictions for at least one year prior to undergoing hormone replacement therapy (HRT). Given the dramatically high incidence of male-to-female transsexuals, who not only fail, but fail miserably to benefit from invasive transsexual procedures post-operatively, health jurisdictions should order rogue practitioners and third-wave health care providers to cease all medication protocols to their transsexual patients, who seek a GRS procedure, and refer these patients to the centralized gender identity clinics. Health jurisdictions should also remove all personnel at the centralized gender identity clinics, who possess a transsexual background, from their positions of authority over transsexual care and replace them with arm's length clinicians. Finally, health jurisdictions should expand the services of these centralized gender identity clinics to deal with the inevitable influx of new patients.

The challenge is few psychiatrists outside the centralized gender identity clinics know of the disorder or how to recognize the disorder or the disorder's similarities to a behavioral addiction. Given the constant intimidation and harassment that candidates for GRS procedures bring to bear on medical practitioners, journalists and politicians, who fall outside of a prescribed belief system, the gender identity clinics are no longer doing fresh research in the field. These clinics are embattled. New psychiatrists and psychologists are badly needed. Further, renewed research is badly needed to avert more tragedies like that of Randy Pallister, a.k.a., Jennifer Pallister. In fact the Pallister tragedy—which was completely avoidable—cost the life of an otherwise brilliant software developer and Ontario tax payers over five hundred thousand dollars of workman's compensation, due to a complete failure on the part of the medical community, who set aside the Hippocratic oath and simply acquiesced.

Another challenge involves patient acceptance. Men and women, who suffer from a behavioral addiction, underlying transsexualism, do not seek genuinely effective remedies, nor will they accept any. The proportion of the ones, who do, will closely follow men and women with an affected anorexia nervosa.

references

[1] LACE DHALILA AMOUR'S CONCEPT OF GROOVE-A-TUDE

Lace Dhalila Amoury, a public cross-dresser, from Rhode Island, refers to the relationship between sex and attitude as groove-a-tude. Amouris quoted as saying "Sex puts the groove in one's tude."

[2] GENDER TALK BROADCAST #357

In her broadcast, dated April 29, 2002, Nancy Nangeroni elaborates on the relationship between sex and transsexual behavior. Nancy Nangeroni is quoted as saying, "[...] anybody who changes their sex is doing it for sexual reasons. There's just no way that you can change your sex for reasons that are not connected to sex." (17:25) Nangeroni continues, "the notion that one can be transsexual and not do it out of sexual motivations I think is silly because it's missing the obvious." (19:13) "the idea that auto -- excuse me, that gynephilic are doing it for more sexual reasons than [...] androphilic transsexuals is similarly silly." (19:21)

more information
transcript of gender talk broadcast #357
http://www.transgression.com/downloads/29Apr02%20Open%20Systems.DOC

[3] WENZEL, HANNE GRO, DAHL, ALV A.: FEMALE PATHOLOGICAL GAMBLERS—A CRITICAL REVIEW OF THE CLINIC FINDINGS

WENZEL, HANNE GRO, DAHL, ALV A.: (2008) Female Pathological Gamblers—A Critical Review of the Clinic Findings. International Journal of Mental Health & Addiction, January 12, 2008, ISSN: 1557-1874 (Print) 1557-1882 (Online); DOI: 10.1007/s11469-008-9174-0.

Appendix A—Notice of Name Change

Message posted by Shannon_B on 17 May 2003 at 10:53am - IP 64.12.96.199

This is a note to let you know that after much agonizing and wrestling with the question, I have finally reached a decision on the issue of changing my name when I transition.

I used to use the 'fake' name Shanin online. Then I decided that I was more comfortable and felt more authentic using my birth name , Sean . I had intended to keep the name Sean , even after I transition to life as a woman. But for a number of reasons, and after more reading/research and consultation with a number of transsexuals, I have decided that a change of name is a good thing, and it will bring benefits to me beyond merely avoiding the awkwardness that an androgynous first name can bring about.

Changing my name will be an important public signification of the change, to myself, to my family, to my friends and to my colleagues at work. It is a rite of passage that I feel I should experience, and it's said to be an important step in terms of "identity consolidation." Mainly, I want to eliminate the prospects for people to continue to see or identify me as "guy Sean ".

I'm very, very comfortable with this decision, and I hope my online friends will be too, although to be honest, it doesn't matter a bit if they are or not. Unlike a flip-flop of online "nick name s", my change(s) have been part of a process of finding comfort in the name in which I intend to live the rest of my life.

I will be changing this forum account, obtaining new email addresses, and announcing my change of name wherever else it's approrpriate or necessary to do so. More importantly, I will be getting my legal change of name underway very shortly, and will be making this change official.

My new name is: Shannon Elizabeth B****

And no...it has nothing to do with a young actress name d SE...I found out about her after the name s had been chosen. Shannon was always the name I knew I wanted. The middle name Elizabeth was selected in consultation with my older sister, and we found that we both have a bit of an admiration for and fascination with Queen Elizabeth the First of England.

So there you have it. Sean is now Shannon. I would appreciate being referred to only by the name Shannon now. My close coworkers at work are now addressing me in email as Shannon/ Sean ...but soon Sean will be a mere memory.

Thanks!
Shannon

Appendix B—The Anorexic Lounge of Deflected Thinking

The Anorexic lounge of deflected thinking is a virtual space where the members of a mythical community of men and women, who suffer from anorexic nervosa disorder, meet to discuss topics of interest, including medical advances in the area of weight loss, how medical practitioners regard anorexia and weight loss and the political issues surrounding anorexia. In the Anorexic lounge weight control is the fundamental to one's health. Anorexics in the Anorexic lounge speak glowingly of better health through weight control. They point to advances in liposuction and stomach reduction surgery as sacrosanct to good health. They opt for self diagnosis over that of medical doctors, who clearly do not understand the merits of such life saving procedures and nor why one would want to rid themselves of unwanted body fat.

On occasion a member of the Anorexic lounge misconstrues something he or she heard or saw in the newspaper. The individual forwards the misreported fact to the group, who are very receptive to the interpretation. Despite hearing clarification of what is meant, the misreported fact is perpetuated. Soon afterwards, a number of members of the community will advance other information that relies on the misreported fact and the misreported fact takes on a life of its own. As the group collectively reports information that relies on the misreported fact, the momentum behind the misreported fact, becomes so massive, the misreported fact turns into the undisputable truth.

Appendix C—Behavioral Addiction & the Human Brain Reward System

the theory behind behavioral addiction

We feel good when neurons in the reward pathway release a neurotransmitter, called dopamine, into the nucleus accumbens and other areas of the brain.

Neurons in the reward pathway communicate with each other by sending electrical signals down their axons. The signal is passed to the next neuron across a small gap called the synapse. (See figure 2: neurons in the reward pathway)

When Dopamine is release into the synapse and crosses to the next neuron and binds to receptors, we feel a burst of pleasure. Excess dopamine is returned to the sending cell through a reuptake gateway. Other nerve cells release GABA, an inhibitory neurotransmitter that works to prevent the receptor nerve from being over stimulated. (See figure 3: an example of a healthy reward system)

Behavioral addiction sets in when endorphins increase the amount of dopamine in the synapse, heightening the feeling of pleasure, while at the same time endorphins block the reuptake of the dopamine, while at the same time endorphins block the release of the inhibitory neurotransmitter. Under such conditions the brain never feels satisfaction and person has to act out the behavioral addiction more and more feel the same degree of satisfaction. (See figure 4: an example of a faulty reward system)

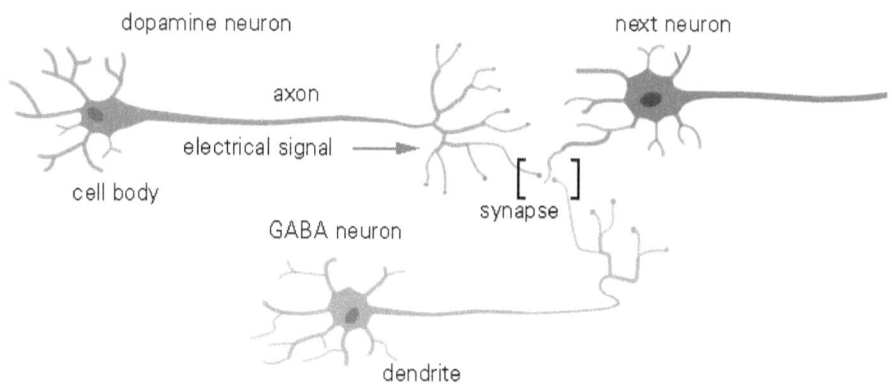

figure 2: neurons in the reward pathway

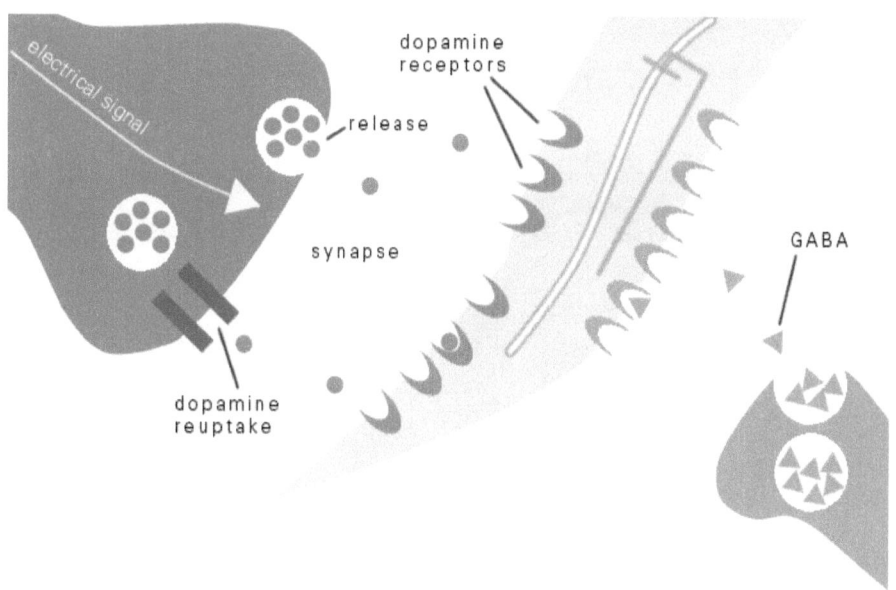

figure 3: an example of a healthy reward system

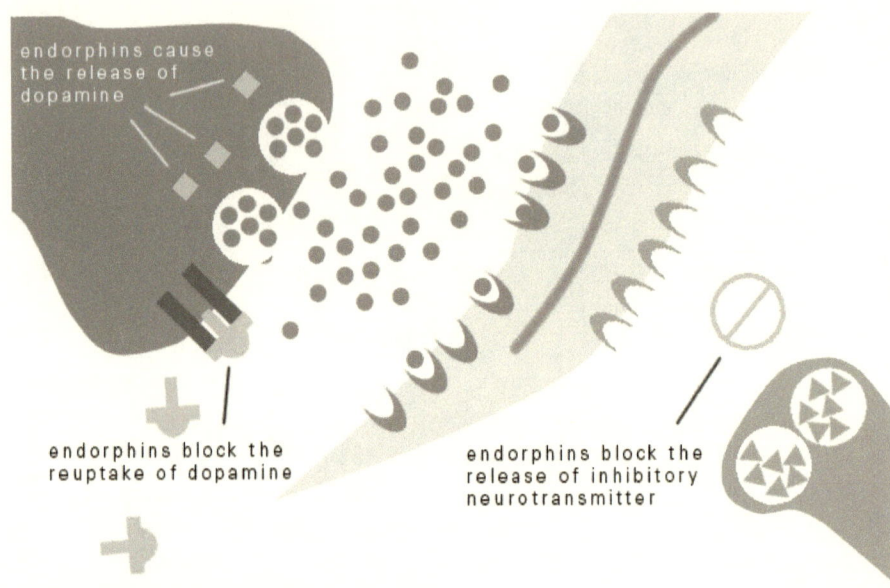

figure 4: an example of a faulty reward system

Appendix D—Transsexualism & the Diagnostic Characteristics of Behavioral Addictions

In a working paper, dated 2006, Grüsser and Thalemann describe the diagnostic characteristics model to distinguish a phenomenon as a behavioral addiction. [1] In virtually all diagnostic characteristics below, transsexualism satisfies the criteria of being a behavioral addiction.

1. *The behavior is exhibited over a long period of time (at least 12 months) in an excessive, aberrant form, deviating from the norm or extravagant (e.g., regarding its frequency and intensity)*
   A transsexual's feelings towards gender transition generally intensify as time progresses. It is extremely rare to see a transsexual, who is committed to a gender transition, veer off. As it is extremely rare to see a transsexual veer off, transsexuals easily surpass twelve months where the behavior becomes more extreme. Ironically, in the vast majority of cases of male-to-female transsexuals, transsexualism was barely on the radar screen prior to the onset of their cross-gender variant behavior. Transsexualism satisfies this diagnostic characteristic of being a behavioral addiction.

2. *Loss of control over the excessive behavior (duration, frequency, intensity, risk) when the behavior started*
   Transsexuals, who begin a gender transition, have difficulty turning off their intense feeling towards gender transition. Their personal gender transition emerges in virtually all aspects of their life. Transsexuals, also, generally fail to appreciate the risk they are taking in the pursuit of in a gender transition. In the case of transsexuals, who are gainfully employed, transsexuals generally do not weigh the issues surrounding re-entering the workforce, should they lose their jobs for one reason or another. Transsexualism satisfies this diagnostic characteristic of being a behavioral addiction.

3. *Reward effect (the excessive behavior is instantly considered to be rewarding)*
   As the person takes steps in their respective gender transition, each one seems the correct one. As the person regards each step in their respective gender transition as being well founded, he/she experiences reward. Over time

the individual has to act out in ever increasing degrees to feel the same degree of satisfaction. Transsexualism satisfies this diagnostic characteristic of being a behavioral addiction.

4.   *Development of tolerance (the behavior is conducted longer, more often and more intensively in order to achieve the desired effect; in unvaried form, intensity and frequency the desired effect fails to appear)*
As a transsexual progresses through a change in gender appearance, he/she often opts to undergo more procedures and not less, indicating the degree of satisfaction derived from each procedure diminishes in intensity, requiring the individual to act out in ever increasing degrees. Transsexualism satisfies this diagnostic characteristic of being a behavioral addiction.

5.   *The behavior that was initially perceived as pleasant, positive and rewarding is increasingly considered to be unpleasant in the course of the addiction*
Virtually all transsexuals publicly cross-dress prior to undergoing a gender transition. The vast majority of transsexuals report their former cross-dressing experiences as being fun and exciting, However, as transsexuals advance in their respective gender transitions, the thought of periodic cross-dressing is abhorrent. Transsexualism satisfies this diagnostic characteristic of being a behavioral addiction.

6.   *Irresistible urge/craving to execute the behavior*
The drivers, underlying transsexualism, are psychological and sexual in nature. Sexual urges are irrepressible. Transsexualism satisfies this diagnostic characteristic of being a behavioral addiction.

7.   *Function (the behavior is primarily employed as a way to regulate emotions/mood)*
The vast majority of transsexuals cannot emotionally function if precluded from acting out cross-gender behavior due to one circumstance or another. Transsexualism satisfies this diagnostic characteristic of being a behavioral addiction.

8.   *Expectancy of effect (expectancy of pleasant/positive effects by carrying out the excessive behavior)*
As a transsexual takes steps in his/her gender transition each one seems the correct one. Transsexuals often report their life 'improving' in the period leading up to a major surgical procedure, which matches their expectations. Transsexualism satisfies this diagnostic characteristic of being a behavioral addiction.

9.   *Limited pattern of behavior (also applies to buildup and follow-up activities)*
See 10.

10. *Cognitive occupation with the build-up, execution and follow-up activities of the excessive behavior and possibly the anticipated effects of the excessively executed behavior*
Transsexuals find gender transition and the associated steps—which is a process and not a state of existence—exciting. Transsexualism satisfies this diagnostic characteristic of being a behavioral addiction.

11. *Irrational, contorted perception of different aspects of the excessive behavior*
The vast majority of transsexuals are utterly incapable of realizing their masturbation thoughts are not indigenous to those of a member of the opposite sex. Further, the vast majority of transsexuals have no real construct of what it means to be a member of the opposite sex. Transsexualism satisfies this diagnostic characteristic of being a behavioral addiction.

12. *Withdrawal symptoms (psychological and physical)*
When circumstances preclude transsexuals from acting out cross-gender variant behavior, they report discomfort and psychological withdrawals. Ironically, the vast majority of transsexuals report that they had no difficult living as members of their assigned gender prior to the consideration of a change in sex. Transsexualism satisfies this diagnostic characteristic of being a behavioral addiction.

13. *Continued execution of the excessive behavior despite negative consequences (health-related, occupational, social)*
On the lose of employment or similar crisis, transsexuals typically do not opt to revert to living as a member of their assigned gender role even on circumstances where doing so would be in their best interest. Transsexualism satisfies this diagnostic characteristic of being a behavioral addiction.

14. *Conditioned / learned reactions (resulting from the confrontation with internal and external stimuli associated with the excessive behavior as well as from cognitive occupation with the excessive behavior)*
Transsexuals are very resistant to pleas from friends, family members and intimacy partners to discontinue what are often ill conceived gender transitions and consider alternative treatments. They often seize on the concept that gender dysphoria is an accurate portray of their feelings and experiences. They also seize on the prescribed treatment regiments for gender dysphoria as set down by the Harry Benjamin International Gender Dysphoria Association (HBIGDA), including hormone replacement therapy (HRT) and genital surgery. As gender identity disorder in sexually developed individuals is fundamentally flawed, so too is the prescribed treatment regiment set down by the HBIGDA. Transsexualism (coupled with the currently flawed medical response) satisfies this diagnostic characteristic of being a behavioral addiction.

15. *Suffering (desire to alleviate perceived suffering)*
The vast majority of male-to-female transsexuals regard any obstruction to

surgical procedures that aid in their respective gender transitions as a form of suffering. A number of male-to-female transsexuals regard gender reassignment surgical (GRS) procedures as 'life saving' surgery and sacrosanct to their health interests, indicating a degree of discomfort in the absence of the procedure. Further, a number of transsexuals report they will commit suicide, if they are unsuccessful with respect to being approved to undergo a GRS procedure or similar procedure that protracts the testes. Male-to-female transsexuals rarely commit suicide in the period leading up to a major surgical procedure that protracts the testes. In fact the vast majority of cases of transsexuals who commit suicide do so after undergoing a GRS procedure and not prior. 5% of all post-operative, male-to-female transsexuals commit suicide, which is 50 times higher than that of the greater population. In the vast majority of cases of suicide the reasons had nothing to do with prejudice. Whether the individual suffers or not is immaterial. The perception is all that matters. Transsexualism satisfies this diagnostic characteristic of being a behavioral addiction.

Transsexualism satisfies virtually all of the diagnostic characteristics of a behavioral addiction. In the case of all sexually developed individuals transsexualism is a behavioral addiction.

references

[1] GRÜSSER THALEMANN DIAGNOSTIC CHARACTERISTICS MODEL FOR IDENTIFY BEHAVIORAL ADDICTIONS IN PHENOMENON

Grüsser SM, Thalemann CN. Verhaltenssucht- Diagnostik, Therapie, Forschung. Bern: Huber; 2006.
more information
visit the site: Diagnostic Instruments for behavioral addiction: an overview
http://www.egms.de/pdf/journals/psm/2007-4/psm000043.pdf

Appendix E—Maria: A Case Study in Post-Operative Regret

Many male-to-female transsexuals have the dream of full transition, yet some find that following surgery the spirit does not fly like the butterfly thought to be cocooning within the male body.

Is the human spirit damaged because of being trapped in the wrong gender's body? Is the male-to-female transsexual conflicted between her male socialization and the difficulties of a woman's lot in these times? Is she destined to continue to experience being treated by others as a male-to-female transsexual rather than as a `real' woman? Will she ever be free of the experience of being different with society's judgments of different as a right or a wrong?

Maria (not 'her' real name) lived for 31 years as a male. Established herself as an accountant with a new firm and in a new city began her first year living full-time as a woman. She was very excited about her upcoming sexual reassignment surgery and the completion of a life dream to become the woman she had always felt she was within this male body. Maria had experienced depressive episodes throughout her life. Previous therapy had her considering this inner emotional pain was connected to being forced physically to be a male.

She would speak of the happiness to come after so many years. To finally be rid of the detested male genitalia would bring a life dream.

The office she worked at knew of her as Maria, but her professional certifications still read Martin—a situation that resulted in her holding a position below her male qualifications. The lower status and pay than he had been use to were rationalized as 'a right of passage' into womanhood. After all, as she would say, "women are often still paid less for the same job as men." Conflict in the work place was experienced as superiors noted her skills and assigned challenging projects with only the promise of better pay and lots of complements.

Maria felt used professionally, conflicted surrounding her gratitude at being employed yet not able to legally sign off her work with professional credentials Martin possessed. She questioned discrimination by her employers and was despondent regarding what she called her new suffering with not benefiting from previous established professional status.

Maria underwent reassignment surgery and legally changed her name, and continued to work with what turned out to be false promises. Seems the employer was only interested in the product at a cheaper rate than Maria deserved. Employment search with a large adult time gap on the resume (namely the time as Martin versus the year as Maria) left her seemingly with an experience similar to a woman return to work following a lengthy period as mommy ...out of date in employers' minds.

During this major step on her journey, Maria attempted to keep her estranged spouse happy by always attempting to appear male for the visits with their 4-year-old daughter. To say that she was male with even breasts tied down was increasingly difficult. Even her daughter would ask about the bulk at the chest when she was hugged by whom she knew as daddy.

Maria today, dresses as an androgynous person, she works as male Martin, the painful issues resulting from her wife's demands that her daughter always have a daddy and the intolerance of the glass ceiling professional have resulted in her hiding other than during intimate moments her gender self. She has decided with the lack of sexual intimacy with another woman she must be a lesbian. She feels she is accepted neither as a lesbian nor as a heterosexual female. Would she have it all reversed? Her answer is a hesitant yes as she feels in private that the body looks as a female ...however she has not become the beautiful butterfly so dearly longed for. Society does not in her experience see her as a lovely creature, tender and beautiful. She does experience financial decline and has an existence similar to many biological females, even in these times of not being given the business/professional benefits of her male counterparts.

Is she happy with her transition, well Maria is considering reverse surgery, even though the costs are very high and the possible minimal results. Will she ever find the happiness she thought would be there with full transition? Doubtful as seemingly the life of this male-to-female transsexual did not turn out to be the release from the cocoon.

www.ingramcontent.com/pod-product-compliance
Lightning Source LLC
Chambersburg PA
CBHW031944170526
45157CB00002B/388